Holt Mathematics

Chapter 5 Resource Book

HOLT, RINEHART AND WINSTON

A Harcourt Education Company

Orlando • Austin • New York • San Diego • London

Copyright © by Holt, Rinehart and Winston

All rights reserved. No part of this publication may be reproduced or transmitted in any form or by any means, electronic or mechanical, including photocopy, recording, or any information storage and retrieval system, without permission in writing from the publisher.

Teachers using HOLT MATHEMATICS may photocopy complete pages in sufficient quantities for classroom use only and not for resale.

Printed in the United States of America

If you have received these materials as examination copies free of charge, Holt, Rinehart and Winston retains title to the materials and they may not be resold. Resale of examination copies is strictly prohibited and is illegal.

Possession of this publication in print format does not entitle users to convert this publication, or any portion of it, into electronic format.

ISBN 0-03-078301-1

6 7   170   09 08

CONTENTS

Blackline Masters

Parent Letter	1
Lesson 5-1 Practice A, B, C	3
Lesson 5-1 Reteach	6
Lesson 5-1 Challenge	8
Lesson 5-1 Problem Solving	9
Lesson 5-1 Reading Strategies	10
Lesson 5-1 Puzzles, Twisters & Teasers	11
Lesson 5-2 Practice A, B, C	12
Lesson 5-2 Reteach	15
Lesson 5-2 Challenge	16
Lesson 5-2 Problem Solving	17
Lesson 5-2 Reading Strategies	18
Lesson 5-2 Puzzles, Twisters & Teasers	19
Lesson 5-3 Practice A, B, C	20
Lesson 5-3 Reteach	23
Lesson 5-3 Challenge	24
Lesson 5-3 Problem Solving	25
Lesson 5-3 Reading Strategies	26
Lesson 5-3 Puzzles, Twisters & Teasers	27
Lesson 5-4 Practice A, B, C	28
Lesson 5-4 Reteach	31
Lesson 5-4 Challenge	32
Lesson 5-4 Problem Solving	33
Lesson 5-4 Reading Strategies	34
Lesson 5-4 Puzzles, Twisters, & Teasers	35
Lesson 5-5 Practice A, B, C	36
Lesson 5-5 Reteach	39
Lesson 5-5 Challenge	41
Lesson 5-5 Problem Solving	42
Lesson 5-5 Reading Strategies	43
Lesson 5-5 Puzzles, Twisters & Teasers	44
Lesson 5-6 Practice A, B, C	45
Lesson 5-6 Reteach	48
Lesson 5-6 Challenge	49
Lesson 5-6 Problem Solving	50
Lesson 5-6 Reading Strategies	51
Lesson 5-6 Puzzles, Twisters & Teasers	52
Lesson 5-7 Practice A, B, C	53
Lesson 5-7 Reteach	56
Lesson 5-7 Challenge	57
Lesson 5-7 Problem Solving	58
Lesson 5-7 Reading Strategies	59
Lesson 5-7 Puzzles, Twisters & Teasers	60
Lesson 5-7 Practice A, B, C	61
Lesson 5-7 Reteach	64
Lesson 5-7 Challenge	65
Lesson 5-7 Problem Solving	66
Lesson 5-7 Reading Strategies	67
Lesson 5-7 Puzzles, Twisters & Teasers	68
Lesson 5-7 Practice A, B, C	69
Lesson 5-7 Reteach	72
Lesson 5-7 Challenge	73
Lesson 5-7 Problem Solving	74
Lesson 5-7 Reading Strategies	75
Lesson 5-7 Puzzles, Twisters & Teasers	76
Answers to Blackline Masters	77

Date _____

Dear Family,

In this chapter, your child will learn about ratios and rates, slope and rates of change, proportions, similar figures, and scale drawings. Your child will also use and convert among customary units of measure.

Proportional reasoning is one of the most important concepts for future work with algebra. Rates and ratios are used in everyday communication and are key to work in architecture, engineering, and all areas of science.

Your child will learn that a **ratio** is a way of comparing two numbers. Survey results are often reported as ratios, such as: "For every 3 students who walk to school, 8 ride their bikes." So, the ratio of walkers to bikers is 3 to 8. The ratio of 3 to 8 can also be written as: $\frac{3}{8}$ or 3:8.

Your child will also learn about rates and how they are used in daily life. A **rate** is a special kind of ratio that compares two amounts with different units. Prices are the most commonly used rates. $1.67 per gallon and 23 cents per ounce are examples. Population density is reported as a rate. An example is 100 people per square mile.

A **unit rate,** or a comparison to *one* of something, is most often used. For instance, car owners are concerned about fuel efficiency—the number of miles a car will travel on *one* gallon.

If Ms. Garcia drove 75 miles while using 3 gallons of gasoline, her rate of gasoline use would be 75 miles per 3 gallons. Dividing each number by 3 will compute the number of miles driven on 1 gallon.

$$\frac{75 \div 3}{3 \div 3} = \frac{25}{1} \quad \frac{25 \text{ miles}}{1 \text{ gallon}}$$

Since the comparison is to *one* gallon, this is a *unit* rate.

Your child will relate rates to graphs. When a graph is a line, a rate called slope measures the steepness of the line. The slope of a line is the ratio of *rise* to *run*.

Slope $= \frac{\text{Rise}}{\text{Run}} = \frac{3}{2}$

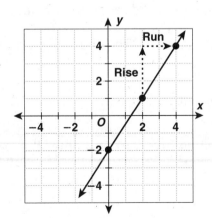

Holt Mathematics

A line such as the one on page 1 has a **constant rate of change**, which means that it changes the same amount during equal intervals. A graph that is not a line has a **variable rate of change**, which means that it changes a different amount during equal intervals.

A **proportion** is a statement that two ratios are equal.

$$\frac{3}{4} = \frac{18}{24}$$

There is a rule that verifies if two ratios are equal. It is called the **cross products rule.** We multiply diagonally across the equal sign: 3×24 and 4×18.

$$\frac{3}{4} \rightleftarrows \frac{18}{24}$$

Since both products are 72, the two ratios form a proportion.

Your child will learn to solve proportions that have variables by using the cross products rule.

Proportions also extend into geometry. Your child will learn about **similar figures.** These are geometric figures that have the same shape and whose corresponding sides are in proportion. To find out if two triangles are similar, use cross products to determine whether the ratios of their corresponding sides are in proportion.

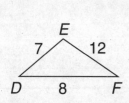
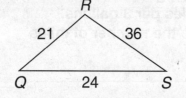

$\frac{DE}{QR} \stackrel{?}{=} \frac{EF}{RS} \stackrel{?}{=} \frac{DF}{QS}$ *Write possible proportions.*

$\frac{7}{21} \stackrel{?}{=} \frac{12}{36} \stackrel{?}{=} \frac{8}{24}$ *Substitute values of lengths and sides.*

$7 \cdot 36 \stackrel{?}{=} 21 \cdot 12$ $12 \cdot 24 \stackrel{?}{=} 36 \cdot 8$ *Find the cross products.*

$252 = 252$ $288 = 288$

Since the cross products are equal, the sides are proportional and the triangles are similar.

For additional resources, visit go.hrw.com and enter the keyword MR7 Parent.

Name _____ Date _____ Class _____

Practice A
5-1 Ratios

Match the ratios.

A farmer has 5 pigs, 13 chickens, and 8 cows.

1. cows to pigs   13:5
2. chickens to pigs   5:8
3. cows to chickens   8:5
4. pigs to cows   8:13

The school orchestra has 9 cellos, 14 flutes, and 17 violins.

5. cellos to violins   9 to 14
6. flutes to cellos   9 to 17
7. violins to flutes   17 to 14
8. cellos to flutes   14 to 9

Miguel has 8 pennies, 5 nickels, and 3 quarters.

9. nickels to pennies
10. pennies to quarters
11. nickels to quarters
12. quarters to pennies

$\frac{3}{8}$
$\frac{5}{3}$
$\frac{5}{8}$
$\frac{8}{3}$

A bowl has 16 grapes, 7 cherries, and 9 strawberries.

13. grapes to strawberries   7:16
14. cherries to grapes   7:9
15. strawberries to cherries 16:9
16. cherries to strawberries 9:7

A baseball team has 4 pitchers, 10 outfielders, and 12 infielders. Write each ratio in all three forms.

17. pitchers to infielders

18. infielders to outfielders

19. pitchers to outfielders

20. outfielders to entire team

21. Meg used 24 red tiles and 12 yellow tiles to make a design. Write the ratio of red tiles to yellow tiles in simplest form.

22. Tell which club has the greater ratio of girls to boys.

	Movie Club	Art Club
Girls	16	8
Boys	14	4

Name _____ Date _____ Class _____

LESSON 5-1 Practice B
Ratios

The annual dog show has 22 collies, 28 boxers, and 18 poodles. Write each ratio in all three forms.

1. collies to poodles

2. boxers to collies

3. poodles to boxers

4. poodles to collies

The Franklin School District has 15 art teachers, 27 math teachers, and 18 Spanish teachers. Write the given ratio in all three forms.

5. art teachers to math teachers

6. math teachers to Spanish teachers

7. Spanish teachers to all teachers

8. art and math teachers to Spanish teachers

9. Thirty-two students are asked whether the school day should be longer. Twenty-four vote "no" and 8 vote "yes." Write the ratio of "no" votes to "yes" votes in simplest form.

10. A train car has 64 seats. There are 48 passengers on the train. Write the ratio of seats to passengers in simplest form.

11. Tell whose CD collection has the greater ratio of rock CDs to total CDs.

	Glen	Nina
Classical CDs	4	8
Rock CDs	9	12
Other CDs	5	7

Copyright © by Holt, Rinehart and Winston.
All rights reserved.

Holt Mathematics

Name _____ Date _____ Class _____

Practice C
LESSON 5-1 Ratios

A traveling theater company has 51 actors, 63 stagehands, and 27 set designers. Write each ratio in all three forms.

1. stagehands to set designers

2. actors to stagehands

3. set designers to actors

4. set designers to stagehands and actors

There are 18 monkeys, 6 gorillas, and 15 other apes in the Primate House at the zoo. Write different combinations of animals that have the following ratios.

5. 2:5 _____

6. 6:5 _____

7. 1:3 _____

8. 6:7 _____

9. 5:13 _____

10. 8:5 _____

11. The class library has 7 French books, 11 Spanish books, 19 art books, 3 Italian books, and 14 science books. What is the ratio of foreign language books to all books in simplest form? _____

12. A market research company tests new commercials. Out of 96 people, 80 people prefer Commercial A and the rest prefer Commercial B. Write the ratio in simplest form of people who prefer Commercial A to Commercial B.

_____

13. Which grade has the greatest ratio of the number of students against wearing a uniform to the total number of students in that grade?

_____

School Uniform Survey			
	6th Grade	7th Grade	8th Grade
For	16	9	5
Against	20	24	21
No Opinion	4	3	1

LESSON 5-1 Reteach
Ratios

A **ratio** is a comparison of two numbers.

Tamara has 2 dogs and 8 fish. The ratio of dogs to fish can be written in three different ways.

	Ratio	Ratio in simplest form
• using the word *to*	2 to 8	1 to 4
• using a colon (:)	2:8	1:4
• writing a fraction	$\frac{2}{8}$	$\frac{1}{4}$

You can read the ratios as 2 to 8 or 1 to 4.

In a basket of fruit, there are 8 apples, 3 bananas, and 5 oranges. Write each ratio in all three forms.

1. apples to bananas

   There are ___ apples and ___ bananas. So, the ratio of apples to bananas is ___ to ___, or ___ : ___, or —.

2. oranges to apples

   There are ___ oranges and ___ apples. So, the ratio of oranges to apples is ___ to ___, or ___ : ___, or —.

3. bananas to oranges

   There are ___ bananas and ___ oranges. So, the ratio of bananas to oranges is ___ to ___, or ___ : ___, or —.

4. apples to all pieces of fruit

   There are ___ apples and ___ pieces of fruit in all. So, the ratio of apples to all pieces of fruit is ___ to ___, or ___ : ___, or —.

A large bouquet of flowers is made up of 18 roses, 16 daisies, and 24 iris. Write each ratio in all three forms.

5. roses to iris

   _____

6. daisies to roses

   _____

7. iris to daisies

   _____

8. roses to all flowers

   _____

Name _____ Date _____ Class _____

LESSON 5-1 Reteach
Ratios (continued)

To compare ratios, write them as fractions with a common denominator. Then compare the numerators.

Tell whose bag has the greater ratio of solid marbles to striped marbles.

	Ken	Val
Solid	5	7
Striped	9	12

Step 1: Write the ratio of solid marbles to striped marbles for each bag. Write each ratio as a fraction.

Ken's Bag: $\frac{\text{Solid}}{\text{Striped}} = \frac{5}{9}$

Val's Bag: $\frac{\text{Solid}}{\text{Striped}} = \frac{7}{12}$

Step 2: Choose a common denominator.

36 is a common denominator for 9 and 12.

Step 3: Write each fraction using the common denominator.

$\frac{5}{9} = \frac{5 \times 4}{9 \times 4} = \frac{20}{36}$ $\frac{7}{12} = \frac{7 \times 3}{12 \times 3} = \frac{21}{36}$

Step 4: Compare the numerators. $20 < 21$

Since $20 < 21$, $\frac{20}{36} < \frac{21}{36}$ and $\frac{5}{9} < \frac{7}{12}$. So, Val's bag has the greater ratio of solid marbles to striped marbles.

9. Tell whose bookshelf has the greater ratio of novels to biographies.

	Tina	Mark
Novels	3	5
Biographies	5	8

For Tina's bookshelf: $\frac{\text{Novels}}{\text{Biographies}} = \underline{}$

For Marks's bookshelf: $\frac{\text{Novels}}{\text{Biographies}} = \underline{}$

Write the fractions so they have a common denominator. Then compare the numerators.

_____'s bookshelf has the greater ratio of novels to biographies.

Name _____ Date _____ Class _____

LESSON 5-1 Challenge
The Golden Ratio

The **Golden Ratio**, which as also known as the Golden Section, was given its name because architects and artists throughout history have used it to produce shapes that are pleasing to look at. They have used this ratio as the ratio of length to width in various shapes.

The Golden Ratio can be represented by a line segment that meets these conditions:
• The line segment is divided into two sections of different lengths.
• The ratio of the longer section to the shorter section is equal to the ratio of the whole segment to the longer section

$$\frac{\text{longer?}(AB)}{\text{shorter?}(BC)} = \frac{\text{whole?}(AC)}{\text{longer?}(AB)}$$

Find each measurement to the nearest millimeter. Use your measurements to write the given ratio. Then write the ratio as a decimal rounded to the nearest hundredth.

1. •————————————•————————————•
   A   B   C

   $\frac{AB}{BC} = \underline{} = \underline{}$ $\frac{AC}{AB} = \underline{} = \underline{}$

2. •————————————•————•
   A   B   C

   $\frac{AB}{BC} = \underline{} = \underline{}$ $\frac{AC}{AB} = \underline{} = \underline{}$

3. •————————•————————————•
   A   B   C

   $\frac{AB}{BC} = \underline{} = \underline{}$ $\frac{AC}{AB} = \underline{} = \underline{}$

4. In which exercise is the line segment divided so that it is closest to representing the golden ratio? Explain.

_____

_____

Problem Solving
5-1 Ratios

Write the correct answer.

1. The Rockport Diner has 8 seats at the counter and 32 seats at tables. Of these seats, 16 are taken. Write the ratio of seats taken to empty seats in simplest form three ways.

2. During the 2001 WNBA season, the Los Angeles Sparks had 28 wins and only 4 losses. Write the ratio of wins to games played in simplest form three ways.

3. For every 300 people surveyed in 2002, 186 said their favorite Winter Olympic sport was figure skating. Write this ratio in simplest form three ways.

4. In 2004, George W. Bush received 286 electoral votes, and John Kerry received 251, and 1 elector voted for John Edwards. Write the ratio of Bush's electoral votes to total electoral votes in simplest form three ways.

Choose the letter for the best answer.

5. There are 62 girls in the seventh grade and 58 boys in the eighth grade. Each grade has 120 students. Which statement correctly compares the ratios of boys to girls in each grade?
   A The eighth-grade ratio is greater.
   B The seventh-grade ratio is greater.
   C The eighth-grade ratio is lesser.
   D Both ratios are equal.

6. Matt has 6 video racing games and 8 video sports games. Which ratio is the ratio of racing games to total video games in simplest form?
   F $\frac{3}{4}$ H $\frac{4}{3}$
   G $\frac{3}{7}$ J $\frac{4}{7}$

7. Which player has the greatest ratio of baskets to total shots?
   A Marisol
   B Nina
   C Joanne
   D Talia

	Baskets	Missed Shots
Marisol	8	8
Nina	7	5
Joanne	2	4
Talia	5	3

LESSON 5-1 Reading Strategies
Build Vocabulary

A **ratio** is a way to compare two numbers.

● ● ● ○ ○

The ratio of black counters to white counters is three to two.

There are three ways to write ratios:

3 to 2   ← read "3 to 2"
3:2   ← read "3 to 2"
$\frac{3}{2}$ ← read "3 to 2"

Now compare the white counters to the black counters.

1. Use a fraction to write this ratio.

2. Write the ratio two other ways.

A team has 5 boys and 4 girls. Use this information to complete Exercises 3 – 6.

3. Write a ratio as a fraction to compare the number of boys to the number of girls.

4. Write this ratio two other ways.

5. Use division to compare the number of girls to the total number of players on the team.

6. Write this ratio two other ways.

Name _____ Date _____ Class _____

Puzzles, Twisters & Teasers
LESSON 5-1 *Find Your Ratio*

Use the letters in the word RATIO to write a fraction for each ratio in simplest form.

RATIO

1. number of Ts:number of Rs _____

2. consonants:vowels _____

3. vowels:total number of letters _____

4. consonants:total number of letters _____

Next, use the letters in the word MATHEMATICS to write the fraction for each ratio in simplest form.

5. consonants:total number of letters _____

6. vowels:consonants _____

7. number of Ms:total number of letters _____

8. number of Hs:total number of letters _____

9. number of Ms and Ts:total number of letters _____

Fraction Box

A = $\frac{1}{7}$
D = $\frac{3}{5}$
E = $\frac{2}{5}$
G = 1
H = $\frac{1}{11}$
I = $\frac{2}{11}$
L = $\frac{2}{7}$
M = $\frac{5}{7}$
O = $\frac{2}{3}$
R = $\frac{6}{7}$
S = $\frac{4}{7}$
T = $\frac{4}{11}$
Y = $\frac{7}{11}$

Find the letters that match your answers in the Fraction Box. To solve the riddle, write the letters on the blanks that correspond to the problem numbers.

What do you need to spot an iceberg 20 miles away?

___ ___ ___ ___
 1   2   2   3

___ ___ ___ ___ ___ ___ ___ ___
 4   5   4   6   7   1   8   9

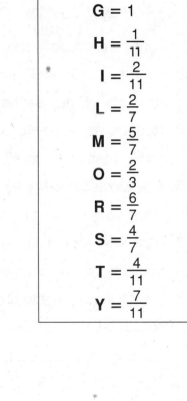

Name _____ Date _____ Class _____

LESSON 5-2
Practice A
Rates

1. To make 2 batches of brownies, Ed needs 4 eggs. How many eggs are needed per batch of brownies?

   $\dfrac{4 \text{ eggs}}{2 \text{ batches}} = \dfrac{\text{eggs}}{1 \text{ batch}}$

   Ed needs _____ eggs to make 1 batch of brownies.

2. Jenny drives 265 miles in 5 hours. What is her average rate of speed in miles per hour?

   $\dfrac{265 \text{ miles}}{5 \text{ hours}} = \dfrac{\text{miles}}{1 \text{ hour}}$

   Jenny's average rate of speed is _____ miles per hour.

3. A job pays $56 for 8 hours of work. How much money does the job pay per hour? _____

4. Ned scores 84 points in 6 games. How many points per game does Ned score? _____

5. A 6-ounce blueberry muffin has 450 calories. How many calories are there per ounce? _____

6. A parking garage charges $21 for 6 hours. How much does the garage charge per hour? _____

7. The Rylands want to drive 360 miles in 8 hours. What should their average speed be in miles per hour?

   _____

8. A plane travels 2,395 miles in 5 hours. What is the plane's average speed? _____

9. A 16-ounce bottle of fruit punch costs $2.40. A 24-ounce bottle of fruit punch costs $3.84. Which size is the better buy?

   $\dfrac{\$2.40}{16 \text{ oz}} = \dfrac{\$}{1 \text{ oz}}$ $\dfrac{\$3.84}{24 \text{ oz}} = \dfrac{\$}{1 \text{ oz}}$

   The _____-ounce bottle costs less per ounce.

   So, the _____-ounce bottle is the better buy.

Name _____ Date _____ Class _____

Practice B
LESSON 5-2 Rates

1. A part-time job pays $237.50 for 25 hours of work. _____
   How much money does the job pay per hour?

2. A class trip consists of 84 students and 6 _____
   teachers. How many students per teacher
   are there?

3. A factory builds 960 cars in 5 days. What is the _____
   average number of cars the factory produces
   per day?

4. The Wireless Cafe charges $5.40 for 45 minutes _____
   of Internet access. How much money does The
   Wireless Cafe charge per minute?

5. A bowler scores 3,152 points in 16 games. _____
   What is his average score in points per game?

6. Melissa drives 238 miles in 5 hours. What is her _____
   average rate of speed?

7. An ocean liner travels 1,233 miles in 36 hours. _____
   What is the ocean liner's average rate of speed?

8. A plane is scheduled to complete a 1,792-mile _____
   flightin 3.5 hours. In order to complete the trip
   on time, what should be the plane's average
   rate of speed?

9. The Nuthouse sells macadamia nuts in three _____
   sizes. The 12 oz jar sells for $8.65, the 16 oz
   jar sells for $10.99, and the 24 oz gift tin costs
   $16.99. Which size is the best buy?

10. Nina paid $37.57 for 13 gallons of gas. Fred paid _____
    $55.67 for 19 gallons of gas. Eleanor paid $48.62
    for 17 gallons of gas. Who got the best buy?

Name _____ Date _____ Class _____

LESSON 5-2 Practice C
Rates

1. Maria earns $603.75 for 35 hours of work. What is her rate of pay per hour? _____

2. The Ranch House serves a 24 oz sirloin steak that has a total of 1,800 calories. How many calories per ounce does the steak have? _____

3. A volunteer stuffs 228 envelopes in an hour. What is the average number of envelopes the volunteer stuffs per minute?
_____

4. A freight train travels 1445 miles in 25 hours. What is the train's average rate of speed? _____

5. June runs 600 yards in 2 min 5 sec. What is her average speed in yards per second? _____

6. A plane travels 294 miles in 45 minutes. What is its average speed in miles per hour? _____

7. A pitcher's earned run average is the number of earned runs allowed per game, with a game defined as 9 innings. A pitcher allows 20 earned runs in 50 innings. What is the pitcher's earned run average? _____

Find each unit price. Then decide which is the better buy.

8. $\frac{\$6.48}{36 \text{ oz}}$ or $\frac{\$8.16}{48 \text{ oz}}$

9. $\frac{\$9.03}{7 \text{ ft}}$ or $\frac{\$15.84}{12 \text{ ft}}$

10. $\frac{\$25.35}{6.5 \text{ lb}}$ or $\frac{\$31.60}{8 \text{ lb}}$

11. $\frac{\$9.16}{0.4 \text{ m}}$ or $\frac{\$13.20}{0.6 \text{ m}}$

Name _____ Date _____ Class _____

LESSON 5-2 Reteach
Rates

A **rate** is a ratio that compares two different kinds of quantities or measurements. Rates can be simplified. Rates sometimes use the words *per* and *for* instead of *to*, such as 55 miles per hour and 3 tickets for $1.

The scale on a map might be 3 inches equals 60 miles. Simplify by dividing both numerator and denominator by the same number.

$$\frac{3 \text{ inches}}{60 \text{ miles}} = \frac{3 \div 3}{60 \div 3} = \frac{1 \text{ inch}}{20 \text{ miles}}, \text{ or 1 inch to 20 miles}$$

A **unit rate** is a rate per 1 unit. So, in a unit rate, the denominator is always 1.

Miguel can type 180 words in 4 minutes.

$$\underbrace{\frac{180 \text{ words}}{4 \text{ minutes}}}_{\text{rate}} = \frac{180 \div 4}{4 \div 4} = \underbrace{\frac{45 \text{ words}}{1 \text{ minute}}}_{\text{unit rate}}, \text{ or } \underbrace{\text{45 words per minute}}_{\text{word form}}$$

Find each unit rate. Write in both fraction and word form.

1. Film costs $7.50 for 3 rolls.

   $$\frac{\$7.50}{3 \text{ rolls}} = \frac{7.50 \div}{3 \div} = \frac{\$}{1 \text{ roll}}$$

   Word form: _____

2. Drive 288 miles on 16 gallons of gas.

   $$\frac{288 \text{ mi}}{16 \text{ gal}} = \frac{288 \div}{16 \div} = \frac{\text{mi}}{1 \text{ gal}}$$

   Word form: _____

3. Earn $52 for 8 hours of work.

   $$\frac{\$52.00}{8 \text{ hr}} = \frac{52 \div}{8 \div} = \frac{\$}{\text{hr}}$$

   Word form: _____

4. Use 5 quarts of water for every 2 pounds of chicken.

   $$\frac{5 \text{ qt}}{2 \text{ lb}} = \frac{5 \div}{2 \div} = \frac{\text{qt}}{\text{lb}}$$

   Word form: _____

5. Snowfall of 12 inches in 4 hours

6. 90 students and 5 teachers

Challenge

LESSON 5-2

The Rate Maze

Contestants A–H must make their way through the maze to the Winners' Circle. To reach the Winners' Circle, each contestant must find a path from his or her current location through sections containing equivalent unit prices or rates. The contestants can move only through sections that share a corner.

Find each unit price or rate. Circle the two contestants who will *not* be able to get to the Winners' Circle.

Name _____ Date _____ Class _____

LESSON 5-2 Problem Solving
Rates

Write the correct answer.

1. A truck driver drives from Cincinnati to Boston in 14 hours. The distance traveled is 840 miles. What is the truck driver's average rate of speed?

2. Melanie earns $97.50 in 6 hours. Earl earns $296.00 in 20 hours. Who earns a higher rate of pay per hour?

3. Mr. Tanney buys a 10-trip train ticket for $82.50. Ms. Elmer buys an unlimited weekly pass for $100 and uses it for 12 trips during the week. Write the unit cost per trip for each person.

4. Metropolitan Middle School has 564 students and 24 teachers. Eastern Middle School has 623 students and 28 teachers. Which school has the lower unit rate of students per teacher?

Choose the letter for the best answer.

5. Which shows 20 pounds per 5 gallons as a unit rate?

   A $\frac{20 \text{ lb}}{1 \text{ gal}}$

   B $\frac{4 \text{ lb}}{1 \text{ gal}}$

   C $\frac{5 \text{ lb}}{1 \text{ gal}}$

   D $\frac{1 \text{ gal}}{4 \text{ lb}}$

6. What is the unit price of a 6-ounce tube of toothpaste that costs $3.75?

   F $0.06
   G $0.23
   H $0.62
   J $0.63

7. Max bought 16 gallons of gas for $40.64. Lydia bought 12 gallons of gas for $31.08. Kesia bought 18 gallons of gas for $45.72. Who got the best buy?

   A Max got the best buy.
   B Lydia got the best buy.
   C Lydia and Kesia both paid the same rate, which is better than Max's rate.
   D Max and Kesia both paid the same rate, which is better than Lydia's rate.

8. A pack of 12 8-ounce bottles of water costs $3.36. What is the unit cost per ounce of bottled water?

   F $0.03 per ounce
   G $0.04 per ounce
   H $0.28 per bottle
   J $0.42 per bottle

Name _____ Date _____ Class _____

LESSON 5-2 Reading Strategies
Build Vocabulary

A **rate** is a special ratio that compares two values that are measured in different units.

- $8 for 2 pounds of beef
  ↓
  $\frac{\$8}{2 \text{ lb}}$
  ↓
  pounds compared to dollars

- 12 miles in 3 hours
  ↓
  $\frac{12 \text{ mi}}{3 \text{ h}}$
  ↓
  miles compared to hours

1. Is the ratio $\frac{5 \text{ hours}}{12 \text{ hours}}$ a rate? Explain.

   _____

2. Is the ratio $\frac{50 \text{ yards}}{18 \text{ seconds}}$ a rate? Explain.

   _____

3. What does the rate $\frac{375 \text{ miles}}{15 \text{ gallons}}$ compare?

   _____

In a **unit rate**, the second quantity in the rate is 1 unit. To write a unit rate, write the rate as a fraction with a denominator of 1.

$$\frac{\$8}{2 \text{ lb}} = \frac{(\$8 \div 2)}{(2 \text{ lb} \div 2)} = \frac{\$4}{1 \text{ lb}}$$
↑ rate   ↑ unit rate

Write *yes* or *no* to tell if each rate is a unit rate. If it is not a unit rate, write the unit rate.

4. $\frac{\$2.75}{1 \text{ h}}$ _____

5. $\frac{100 \text{ mi}}{4 \text{ gal}}$ _____

6. $\frac{35 \text{ lb}}{1 \text{ box}}$ _____

7. $\frac{40 \text{ ft}}{5 \text{ s}}$ _____

Name _____ Date _____ Class _____

Puzzles, Twisters, & Teasers
LESSON 5-2 At This Rate...

Draw a line to connect each rate to the equivalent unit rate.

1. $\frac{\$22.80}{8 \text{ lb}}$ $\frac{18 \text{ ft}}{1 \text{ s}}$ O

2. $\frac{288 \text{ ft}}{16 \text{ s}}$ $\frac{\$2.75}{1 \text{ lb}}$ I

3. $\frac{\$16.50}{6 \text{ lb}}$ $\frac{17.6 \text{ ft}}{1 \text{ s}}$ M

4. $\frac{792 \text{ ft}}{45 \text{ s}}$ $\frac{\$0.05}{1 \text{ min}}$ N

5. $\frac{\$3.74}{68 \text{ min}}$ $\frac{\$2.85}{1 \text{ lb}}$ K

6. $\frac{\$4.05}{75 \text{ min}}$ $\frac{\$0.06}{1 \text{ min}}$ A

Next, choose the best buy. Circle the letter next to your answer.

7. E. $\frac{\$3.33}{4.5 \text{ lb}}$ L. $\frac{\$2.28}{3 \text{ lb}}$ C. $\frac{\$2.92}{4 \text{ lb}}$

8. O. $\frac{\$45.36}{21 \text{ gal}}$ [O]   B. $\frac{\$38.97}{18 \text{ gal}}$ T. $\frac{\$57.25}{25 \text{ gal}}$

9. S. $\frac{\$2.16}{36 \text{ min}}$ A. $\frac{\$3.30}{60 \text{ min}}$ [A]   G. $\frac{\$5.04}{72 \text{ min}}$

10. P. $\frac{\$34.32}{8 \text{ yd}}$ W. $\frac{\$49.92}{12 \text{ yd}}$ R. $\frac{\$65.44}{16 \text{ yd}}$

Write the letters that are next to your answers above the problem numbers to solve the riddle.

Where do tadpoles hang their coats?

__ __ __
3. 6. 5.

__ __ __ __ __ __ __ __ __
7. 10. 2. 9. 1. 10. 2. 8. 4.

Name _____ Date _____ Class _____

LESSON 5-3 Practice A
Slope and Rates of Change

Choose the letter for the best answer.

1. What is the slope of the line?

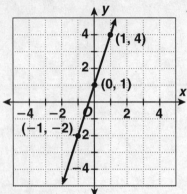

A $\frac{1}{3}$ C $-\frac{1}{3}$
B 3 D -3

2. What is the slope of the line?

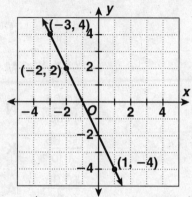

A $\frac{1}{2}$ C $-\frac{1}{2}$
B 2 D -2

Use the given slope and point to graph each line.

3. slope = 2; (1, −2)

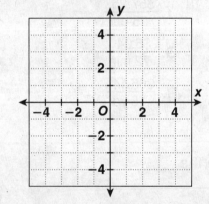

4. slope = −1; (2, 0)

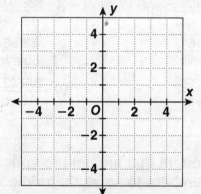

Tell whether each graph shows a constant or variable rate of change.

5.

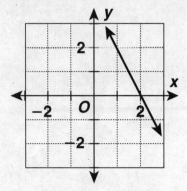

_____

6.

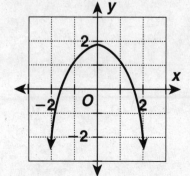

_____

Name _____ Date _____ Class _____

LESSON 5-3 Practice B
Slope and Rates of Change

Tell whether the slope is positive or negative. Then find the slope.

1.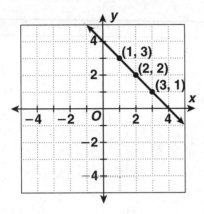

2.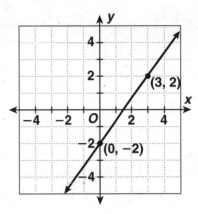

_____   _____

Use the given slope and point to graph each line.

3. $-\frac{1}{2}$; (2, 4)

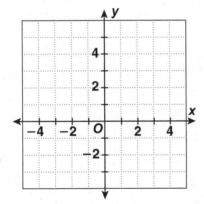

4. $\frac{1}{3}$; (−1, −2)

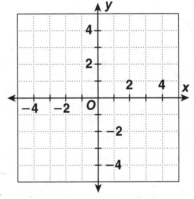

Tell whether each graph shows a constant or variable rate of change.

5.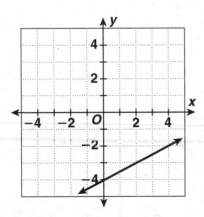

6.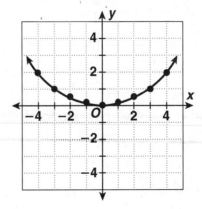

_____   _____

Copyright © by Holt, Rinehart and Winston.
All rights reserved.

Holt Mathematics

Name _____ Date _____ Class _____

LESSON 5-3 Practice C
Slope and Rates of Change

Use the given slope and point to graph each line.

1. $-\frac{1}{3}$; $(-3, 5)$

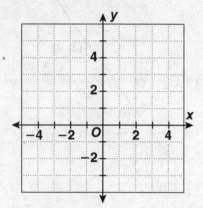

2. $\frac{1}{2}$; $(2, 1)$

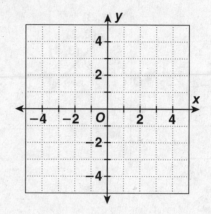

Tell whether each graph shows a constant or variable rate of change.

3.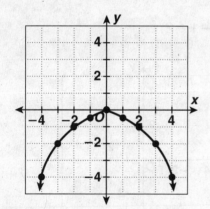

_____

4.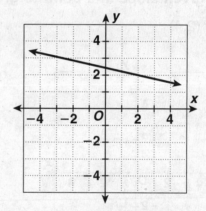

_____

5. The graph at the right shows the cost per pound of buying grapes.

   a. Is the cost per pound a constant or a variable rate?

   _____

   b. What is the cost per pound of grapes?

   _____

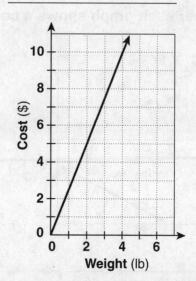

Copyright © by Holt, Rinehart and Winston.
All rights reserved.

Holt Mathematics

Name _____ Date _____ Class _____

LESSON 5-3 Reteach
Slope and Rates of Change

The **slope** of a line is a ratio that measures the steepness of that line. The sign of the slope tells whether the line is rising or falling.

You can find the slope of a line by comparing any two points on that line. Find the slope of the line between (2, 1) and (4, 4).

$$\text{slope} = \frac{\text{rise}}{\text{run}} = \frac{\text{up } (+) \text{ or down } (-)}{\text{right } (+) \text{ or left } (-)}$$

$$= \frac{4-1}{4-2} = \frac{3}{2}$$

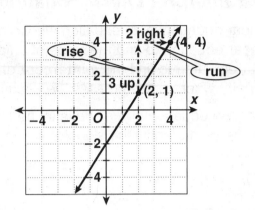

Use the graph to complete the statements.

1. You have to move up ____ and right ____ to go from one point to the other.

   $$\text{slope} = \frac{\text{rise}}{\text{run}} = \frac{\text{up } (+) \text{ or down } (-)}{\text{right } (+) \text{ or left } (-)} = \frac{3}{2}$$

2. The slope of the line is ____.

Find the slope.

3.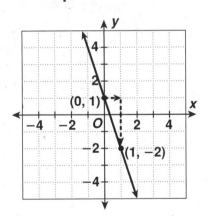

    $\text{slope} = \dfrac{\text{up or down}}{\text{right or left}} = \underline{\quad} = \underline{\quad}$

4.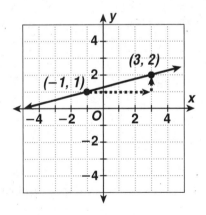

    $\text{slope} = \dfrac{\text{up or down}}{\text{right or left}} = \underline{\quad}$

5. Explain how you can use the point (2, 3) and the slope of 2 to draw a line.

_____

_____

23   Holt Mathematics

Name _____ Date _____ Class _____

LESSON 5-3 Challenge

Interpret Linear Equations

A linear equation is an equation whose graph is a line. You can use a linear equation to find the slope of a line. You can also use a linear equation to find the **y-intercept** of a line. The y-intercept is the point where a line crosses the y-axis.

When the equation of the line is in the form $y = mx + b$, you can use m to identify the slope of the line and b to identify the y-intercept.

$y = mx + b$
$y = -4x + 2$

The slope is -4 and the y-intercept is 2. The coordinates of the y-intercept are $(0, 2)$.

$y = mx + b$
$y = \left(\frac{2}{3}\right)x - 1$

The slope is $\frac{2}{3}$ and the y-intercept is -1. The coordinates of the y-intercept are $(0, -1)$.

Find the slope and the coordinates of the y-intercept for each equation.

1. $y = 8x + 16$

2. $y = -3x + 4$

3. $y = \left(\frac{2}{3}\right)x - 4$

4. $y = -\left(\frac{1}{2}\right)x + \frac{3}{4}$

5. $y = 9x$

6. $y = -5x - 6$

7. $y = -x - 3\frac{1}{3}$

8. $y = \frac{2}{3}x + 4$

9. $y = \left(-\frac{1}{8}\right)x$

Rewrite each equation so that it is in the form $y = mx + b$. Then find the slope and the y-intercept.

10. $y - 1 = 2x + 2$

11. $y + 4 = x + 1$

12. $2y = 2x + 8$

13. $3y = x - 6$

Name _____ Date _____ Class _____

LESSON 5-3 Problem Solving
Slope and Rates of Change

Write the correct answer.

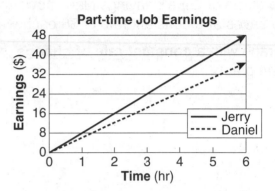
Part-time Job Earnings

1. How much does Jerry earn per hour?

2. What is the slope of the graph that represents Daniel's rate of pay? How much does Daniel earn per hour?

3. Jerry and Daniel each worked 10 hours this week. How much more than Daniel did Jerry earn?

4. For more than 10 hours of work, the rate of pay is 1.5 times that shown in the graph. How much would Jerry earn by working 14 hours?

Choose the letter of the best answer.

The graph shows the changing height of two rockets over time. Use the graph to solve problems 5 and 6.

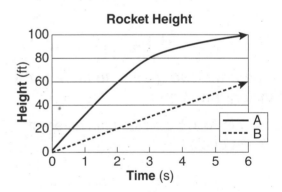
Rocket Height

5. Which statement is true?
   A Both Rocket A and Rocket B have a constant rate of change in height.
   B Both Rocket A and Rocket B have a variable rate of change in height.
   C Rocket A has a variable rate of change in height, but Rocket B does not.
   D Rocket B has a variable rate of change in height, but Rocket A does not.

6. How fast is the height of rocket B increasing?
   F 5 feet per second
   G 10 feet per second
   H 20 feet per second
   J 40 feet per second

7. Jamaal plotted the point (1, −2). Then he used the slope $-\frac{2}{3}$ to find another point on the line. Which point could be the point that Jamaal found?
   A (−1, −1)
   B (4, 0)
   C (2, −3)
   D (4, −4)

Name _____ Date _____ Class _____

LESSON 5-3 Reading Strategies
Compare and Contrast

The graph of Lupe's savings plan shows that she saves at a constant rate of $5 each week.

A graph with a **constant rate of change** is a line graph.

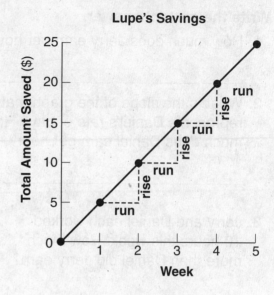

The graph of Alok's plan shows the amount he saves is different from one week to the next.

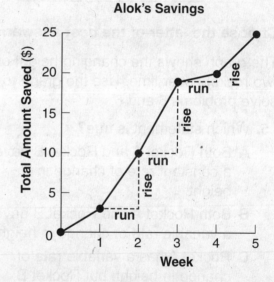

Compare the two graphs to answer the following questions.

1. How is Alok's savings plan different than Lupe's?

   _____

   _____

2. How do the graphs show a difference in their savings plans?

   _____

3. Compare the totals that Alok and Lupe had saved after 4 weeks.

   _____

4. Why is one graph a straight line and the other is not?

   _____

   _____

Copyright © by Holt, Rinehart and Winston.
All rights reserved.

Holt Mathematics

Name _____ Date _____ Class _____

Puzzles, Twisters, & Teasers
LESSON 5-3 *Slippin' and Slopin'!*

Across

4. The _____ of a line is a measure of its steepness and is the ratio of rise to run.

5. A line that slopes upward has a _____ slope.

6. When you know the slope and any one point on a line, you can graph the _____

7. A graph that is a line has a _____ rate of change.

Down

1. A graph that is a curve has a _____ rate of change.

2. The _____ describes the difference between the y-coordinates of two points.

3. A line that slopes downward has a _____ slope.

5. Given one _____ and the slope, you can graph a line.

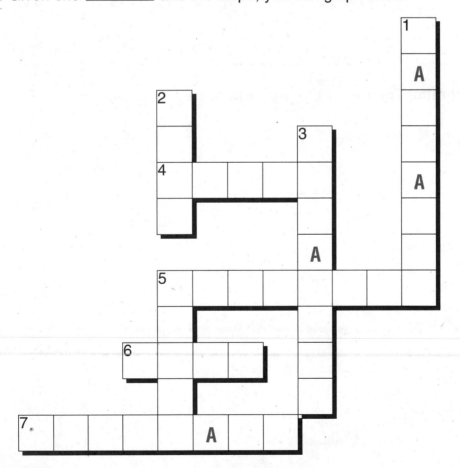

Practice A
5-4 Identifying and Writing Proportions

Write the ratios in simplest form. Determine if the ratios are proportional by comparing them.

1. $\dfrac{1}{4}, \dfrac{3}{12}$

2. $\dfrac{2}{3}, \dfrac{6}{9}$

3. $\dfrac{4}{5}, \dfrac{15}{20}$

4. $\dfrac{3}{6}, \dfrac{6}{12}$

5. $\dfrac{5}{6}, \dfrac{16}{18}$

6. $\dfrac{2}{5}, \dfrac{6}{15}$

7. $\dfrac{1}{3}, \dfrac{3}{9}$

8. $\dfrac{4}{6}, \dfrac{7}{12}$

9. $\dfrac{3}{4}, \dfrac{18}{24}$

10. $\dfrac{2}{3}, \dfrac{9}{15}$

11. $\dfrac{2}{4}, \dfrac{9}{20}$

12. $\dfrac{3}{5}, \dfrac{15}{25}$

Find an equivalent ratio. Then write the proportion.

13. $\dfrac{1}{2}$

14. $\dfrac{3}{4}$

15. $\dfrac{5}{8}$

16. $\dfrac{4}{6}$

17. $\dfrac{1}{7}$

18. $\dfrac{10}{25}$

Practice B
5-4 Identifying and Writing Proportions

Determine whether the ratios are proportional.

1. $\dfrac{3}{4}, \dfrac{24}{32}$

2. $\dfrac{5}{6}, \dfrac{15}{18}$

3. $\dfrac{10}{12}, \dfrac{20}{32}$

4. $\dfrac{7}{10}, \dfrac{22}{30}$

5. $\dfrac{9}{6}, \dfrac{21}{14}$

6. $\dfrac{7}{9}, \dfrac{24}{27}$

7. $\dfrac{4}{10}, \dfrac{6}{15}$

8. $\dfrac{7}{12}, \dfrac{13}{20}$

9. $\dfrac{4}{9}, \dfrac{6}{12}$

10. $\dfrac{7}{8}, \dfrac{14}{16}$

11. $\dfrac{9}{10}, \dfrac{45}{50}$

12. $\dfrac{3}{7}, \dfrac{10}{21}$

Find a ratio equivalent to each ratio. Then use the ratios to write a proportion.

13. $\dfrac{7}{9}$

14. $\dfrac{11}{12}$

15. $\dfrac{14}{15}$

16. $\dfrac{35}{55}$

17. $\dfrac{14}{10}$

18. $\dfrac{25}{18}$

LESSON 5-4 Practice C
Identifying and Writing Proportions

Determine whether the ratios are proportional.

1. $\frac{7}{11}, \frac{42}{60}$

2. $\frac{10}{18}, \frac{38}{72}$

3. $\frac{18}{28}, \frac{27}{42}$

_____ _____ _____

4. $\frac{6}{14}, \frac{15}{35}$

5. $\frac{9}{24}, \frac{16}{40}$

6. $\frac{12}{39}, \frac{20}{65}$

_____ _____ _____

Find a ratio equivalent to each ratio. Then use the ratios to write a proportion.

7. $\frac{7}{31}$

8. $\frac{24}{51}$

9. $\frac{6}{29}$

_____ _____ _____

10. $\frac{14}{23}$

11. $\frac{17}{39}$

12. $\frac{25}{32}$

_____ _____ _____

Complete each table of equivalent ratios.

13. 4 CDs to 10 tapes

CDs	2		10		28
Tapes		10		30	

14. 9 triangles per 6 circles

Triangles		9		30	
Circles	2		8		50

Find two ratios equivilent to each given ratio.

15. 10:21 _____

16. 15:8 _____

17. $\frac{5}{9}$ _____

18. $\frac{24}{11}$ _____

Name _____ Date _____ Class _____

LESSON 5-4 Reteach
Identifying and Writing Proportions

Two ratios that are equal form a **proportion**. To determine whether two ratios are proportional, find the cross products of the ratios. If the cross products are equal, then the ratios are proportional.

If $a \cdot d = b \cdot c$, then $\dfrac{a}{b} = \dfrac{c}{d}$.

Are $\dfrac{6}{9}$ and $\dfrac{8}{12}$ proportional?
Find the cross products.
$6 \cdot 12 = 72$ and $9 \cdot 8 = 72$
Since the cross products are equal, $\dfrac{6}{9}$ and $\dfrac{8}{12}$ are proportional.
So, $\dfrac{6}{9} = \dfrac{8}{12}$.

Are $\dfrac{4}{10}$ and $\dfrac{3}{8}$ proportional?
Find the cross products.
$4 \cdot 8 = 32$ and $10 \cdot 3 = 30$
Since the cross products are not equal, $\dfrac{4}{10}$ and $\dfrac{3}{8}$ are not proportional.
So, $\dfrac{4}{10} \neq \dfrac{3}{8}$.

Find the cross products to determine if the ratios are proportional.

1. $\dfrac{15}{21}, \dfrac{5}{7}$

    $15 \cdot \underline{\quad} = \underline{\quad}$ $21 \cdot \underline{\quad} = \underline{\quad}$

    Are the ratios proportional? _____

2. $\dfrac{6}{9}, \dfrac{9}{15}$

    $6 \cdot \underline{\quad} = \underline{\quad}$ $\underline{\quad} \cdot \underline{\quad} = \underline{\quad}$

    Are the ratios proportional? _____

3. $\dfrac{15}{6}, \dfrac{9}{4}$

4. $\dfrac{12}{24}, \dfrac{5}{10}$

5. $\dfrac{20}{12}, \dfrac{15}{9}$

_____   _____   _____

You can write a proportion from a given ratio. Multiply or divide the numerator and denominator of the ratio by the same number.

$\dfrac{9}{12} = \dfrac{9 \div 3}{12 \div 3} = \dfrac{3}{4}$ So, $\dfrac{9}{12} = \dfrac{3}{4}$. $\dfrac{9}{12} = \dfrac{9 \cdot 4}{12 \cdot 4} = \dfrac{36}{48}$ So, $\dfrac{9}{12} = \dfrac{36}{48}$.

Find an equivalent ratio. Then write the proportion.

6. $\dfrac{6}{10}$

7. $\dfrac{10}{15}$

8. $\dfrac{18}{24}$

_____   _____   _____

Name _____ Date _____ Class _____

LESSON 5-4 Challenge
What's in the Set?

The ratios below describe a set of polygons. Use the ratios to find the number of each type of polygon in the set. All ratios are written in simplest form.

- Equilateral triangles to isosceles triangles is 2 to 3
- Isosceles triangles to scalene triangles is 6:5
- Triangles to rectangles is $\frac{5}{4}$
- Parallelograms to triangles is $\frac{4}{3}$

1. What is the fewest number of each polygon that can be in the set?

    a. Equilateral triangles _____

    b. Isosceles triangles _____

    c. Scalene triangles _____

    d. All triangles _____

    e. Rectangles _____

    f. Parallelograms _____

2. Explain the order in which you determined the number of each polygon in the set.

    _____

    _____

    _____

3. Other sets of polygons with the same ratios are possible. Find the number of polygons in another set that has the same ratios.

    a. Equilateral triangles _____

    b. Isosceles triangles _____

    c. Scalene triangles _____

    d. All triangles _____

    e. Rectangles _____

    f. Parallelograms _____

Copyright © by Holt, Rinehart and Winston.
All rights reserved.

Holt Mathematics

Problem Solving
5-4 Identifying and Writing Proportions

Write the correct answer.

1. Jeremy earns $234 for 36 hours of work. Miguel earns $288 for 40 hours of work. Are the pay rates of these two people proportional? Explain.

2. Marnie bought two picture frames. One is 5 inches by 8 inches. The other is 15 inches by 24 inches. Are the ratios of length to width proportional for these frames? Explain.

3. The ratio of adults to children at a picnic is 4 to 5. The total number of people at the picnic is between 20 and 30. Write an equivalent ratio to find how many adults and children are at the picnic.

4. A recipe for fruit punch calls for 2 cups of pineapple juice for every 3 cups of orange juice. Write an equivalent ratio to find how many cups of pineapple juice should be used with 12 cups of orange juice.

Choose the letter for the best answer.

5. A clothing store stocks 5 blouses for every 3 pairs of pants. Which ratio is proportional for the number of pairs of pants to blouses?
   - A 15:9
   - B 3:8
   - C 12:20
   - D 18:25

6. To make lemonade, you can mix 4 teaspoons of lemonade powder with 16 ounces of water. What is the ratio of powder to water?
   - F 4:32
   - G 32:8
   - H 24:64
   - J 32:128

7. The town library is open 4 days per week. Suppose you use the ratio of days open to days in a week to find the number of days open in 5 weeks. What proportion could you write?

   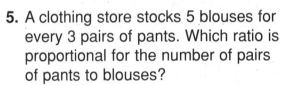

   - A $\frac{4}{7} = \frac{20}{25}$
   - B $\frac{7}{4} = \frac{21}{12}$
   - C $\frac{4}{7} = \frac{20}{28}$
   - D $\frac{4}{7} = \frac{20}{35}$

8. At a factory, the ratio of defective parts to total number of parts is 3:200. Which is an equivalent ratio?
   - F 6:1000
   - G 150:10,000
   - H 30:1000
   - J 1,000:10,000

LESSON 5-4

Reading Strategies
Compare and Contrast

A **proportion** is two equal ratios.
Here are two ratios: $\frac{6}{8}$ and $\frac{9}{12}$.

To find out if they are equal, reduce ratios to simplest form.

$\frac{6}{8} = \frac{3}{4}$ $\frac{9}{12} = \frac{3}{4}$

$\frac{6}{8}$ and $\frac{9}{12}$ are **equal ratios.** They form a proportion.

Read: "6 is to 8 as 9 is to 12."

Compare these two ratios: $\frac{4}{7}$ and $\frac{5}{9}$.

These ratios are in simplest form, but they are *not equal*.

→ $\frac{4}{7} \neq \frac{5}{9}$

$\frac{4}{7}$ and $\frac{5}{9}$ are not equal ratios. They do *not* form a proportion.

Use the ratios $\frac{4}{6}$ and $\frac{1}{3}$ to answer Exercises 1–3.

1. Reduce $\frac{4}{6}$ to simplest form. _____

2. Compare $\frac{4}{6}$ and $\frac{1}{3}$. Are they equal ratios?
   _____

3. Do these two ratios form a proportion? Why or why not?
   _____

Use the ratios $\frac{2}{5}$ and $\frac{4}{10}$ for Exercises 4–6.

4. Reduce $\frac{4}{10}$ to simplest form. _____

5. Compare $\frac{2}{5}$ and $\frac{4}{10}$. Are they equal ratios?
   _____

6. Do $\frac{2}{5}$ and $\frac{4}{10}$ form a proportion? Why or why not?
   _____

Name _____ Date _____ Class _____

Puzzles, Twisters & Teasers
LESSON 5-4 *Draw the Line!*

Draw lines to connect the proportional ratios. Then start with number 1 and find the letter next to the answer. Use the letters with each correct answer to solve the riddle.

1. $\dfrac{8}{14}$ $\dfrac{20}{12}$ **T**

2. $\dfrac{2}{4}$ $\dfrac{3}{9}$ **E**

3. $\dfrac{18}{24}$ $\dfrac{5}{6}$ **R**

4. $\dfrac{10}{6}$ $\dfrac{2}{7}$ **S**

5. $\dfrac{6}{21}$ $\dfrac{2}{10}$ **D**

6. $\dfrac{6}{20}$ $\dfrac{3}{10}$ **Y**

7. $\dfrac{1}{3}$ $\dfrac{25}{15}$ **A**

8. $\dfrac{1}{5}$ $\dfrac{3}{4}$ **I**

9. $\dfrac{5}{3}$ $\dfrac{12}{24}$ **H**

10. $\dfrac{10}{12}$ $\dfrac{4}{7}$ **U**

Why do soccer players do well in math?

__ __ __ __ __ __ __   __ __ __ __ __ __ __ __ __ __ .
4 2 7 6 1 5 7   4 2 7 3 10 2 7 9 8 5

Name _____ Date _____ Class _____

LESSON 5-5 Practice A
Solving Proportions

Find the cross products.

1. $\dfrac{1}{2} = \dfrac{x}{8}$

2. $\dfrac{a}{6} = \dfrac{7}{9}$

3. $\dfrac{5}{b} = \dfrac{8}{10}$

_____ _____ _____

Use cross products to solve each proportion.

4. $\dfrac{2}{5} = \dfrac{x}{10}$

5. $\dfrac{1}{3} = \dfrac{z}{15}$

6. $\dfrac{3}{8} = \dfrac{s}{16}$

_____ _____ _____

7. $\dfrac{4}{r} = \dfrac{1}{4}$

8. $\dfrac{10}{h} = \dfrac{5}{6}$

9. $\dfrac{1}{d} = \dfrac{4}{12}$

_____ _____ _____

10. $\dfrac{w}{9} = \dfrac{6}{18}$

11. $\dfrac{t}{8} = \dfrac{3}{4}$

12. $\dfrac{k}{5} = \dfrac{9}{15}$

_____ _____ _____

13. $\dfrac{3}{6} = \dfrac{1}{f}$

14. $\dfrac{2}{7} = \dfrac{6}{d}$

15. $\dfrac{2}{9} = \dfrac{4}{c}$

_____ _____ _____

16. $\dfrac{a}{20} = \dfrac{15}{10}$

17. $\dfrac{21}{k} = \dfrac{7}{4}$

18. $\dfrac{3}{8} = \dfrac{n}{40}$

_____ _____ _____

19. Yolanda drove 50 miles in 2 hours at a constant speed. Use a proportion to find how long it would take her to drive 150 miles at the same speed.

_____

Practice B
5-5 Solving Proportions

Use cross products to solve each proportion.

1. $\dfrac{2}{5} = \dfrac{x}{35}$
2. $\dfrac{7}{r} = \dfrac{1}{4}$
3. $\dfrac{k}{75} = \dfrac{9}{15}$

4. $\dfrac{1}{3} = \dfrac{z}{27}$
5. $\dfrac{2}{11} = \dfrac{12}{d}$
6. $\dfrac{24}{s} = \dfrac{4}{12}$

7. $\dfrac{w}{42} = \dfrac{6}{7}$
8. $\dfrac{t}{54} = \dfrac{2}{9}$
9. $\dfrac{3}{8} = \dfrac{a}{64}$

10. $\dfrac{17}{34} = \dfrac{7}{f}$
11. $\dfrac{15}{h} = \dfrac{5}{6}$
12. $\dfrac{4}{15} = \dfrac{36}{c}$

13. $\dfrac{z}{25} = \dfrac{12}{5}$
14. $\dfrac{36}{k} = \dfrac{9}{4}$
15. $\dfrac{5}{14} = \dfrac{n}{42}$

16. $\dfrac{8}{9} = \dfrac{40}{m}$
17. $\dfrac{7}{c} = \dfrac{63}{54}$
18. $\dfrac{24}{21} = \dfrac{s}{35}$

19. $\dfrac{e}{22} = \dfrac{6}{15}$
20. $\dfrac{3}{v} = \dfrac{12}{17}$
21. $\dfrac{5}{14} = \dfrac{4}{a}$

22. Eight oranges cost $1.00. How much will 5 dozen oranges cost?

23. A recipe calls for 2 eggs to make 10 pancakes. How many eggs will you need to make 35 pancakes?

Name _____ Date _____ Class _____

LESSON 5-5 Practice C
Solving Proportions

Use cross products to solve each proportion.

1. $\dfrac{3}{7} = \dfrac{x}{49}$

2. $\dfrac{4}{11} = \dfrac{z}{55}$

3. $\dfrac{16}{9} = \dfrac{64}{a}$

4. $\dfrac{13}{r} = \dfrac{1}{5}$

5. $\dfrac{17}{41} = \dfrac{34}{f}$

6. $\dfrac{k}{18} = \dfrac{11}{3}$

7. $\dfrac{w}{39} = \dfrac{7}{13}$

8. $\dfrac{7}{19} = \dfrac{t}{95}$

9. $\dfrac{65}{j} = \dfrac{13}{17}$

10. $\dfrac{15}{h} = \dfrac{5}{17}$

11. $\dfrac{1.7}{3} = \dfrac{d}{21}$

12. $\dfrac{5}{19} = \dfrac{35}{c}$

13. $\dfrac{28}{9} = \dfrac{19.6}{m}$

14. $\dfrac{e}{136} = \dfrac{13}{17}$

15. $\dfrac{3.7}{3} = \dfrac{s}{21}$

Arrange the four numbers to form a proportion that is true.

16. 40, 50, 80, 64

17. 7, 9, 56, 72

18. 50, 45, 15, 150

19. A farmer can harvest 54 pounds of corn from each acre in his field. How much corn can he get from 12.5 acres?

20. A recipe for lasagna calls for 3 pounds of tomatoes to serve 5 people. A caterer wants to make enough lasagna to serve 110 people. How many pounds of tomatoes does he need?

Name _____ Date _____ Class _____

LESSON 5-5 Reteach
Solving Proportions

Solving a proportion is like solving an equation involving fractions.
- Multiply both sides of the equation by the denominator of the fraction containing the variable.
- If the variable is in the denominator, invert both fractions in the proportion.

$$\frac{n}{7} = \frac{20}{28}$$

$$7 \cdot \frac{n}{7} = 7 \cdot \frac{20}{28}$$

$$n = \frac{7 \cdot 20}{28} = \frac{140}{28}$$

$$n = 5$$

$$\frac{12}{x} = \frac{9}{6}$$ *(Invert both fractions.)*

$$\frac{x}{12} = \frac{6}{9}$$

$$12 \cdot \frac{x}{12} = 12 \cdot \frac{6}{9}$$

$$x = \frac{12 \cdot 6}{9} = \frac{72}{9}$$

$$x = 8$$

Solve the proportion.

1. $\frac{a}{2} = \frac{27}{18}$

   ___ $\cdot \frac{a}{2} = \frac{27}{18} \cdot$ ___

   $a = \frac{27 \cdot \_\_}{18}$

   $a = \frac{\_\_}{18}$

   $a = $ ___

2. $\frac{8}{12} = \frac{n}{9}$

   ___ $\cdot \frac{8}{12} = \frac{n}{9} \cdot$ ___

   $\frac{\_\_ \cdot 8}{12} = n$

   $\frac{\_\_}{12} = n$

   $n = $ ___

3. $\frac{10}{t} = \frac{4}{6}$

   $\frac{t}{10} = \frac{6}{4}$

   ___ $\cdot \frac{t}{10} = \frac{6}{4} \cdot$ ___

   $t = \frac{6 \cdot \_\_}{4}$

   $t = \frac{\_\_}{4}$

   $t = $ ___

4. $\frac{x}{15} = \frac{8}{10}$

5. $\frac{7}{3} = \frac{w}{18}$

6. $\frac{3}{2} = \frac{15}{c}$

LESSON 5-5: Reteach
Solving Proportions (continued)

You can use proportions to solve word problems.

A fruit punch is made with 32 ounces of ginger ale for every 12 ounces of frozen orange juice concentrate. How much ginger ale should you use for 30 ounces of orange juice concentrate?

- Set up a proportion comparing the amounts of ginger ale to orange juice concentrate.
- The first ratio shows the given recipe for the fruit punch.
- The second ratio shows the unknown amount of ginger ale as the variable g.
- Then solve the proportion.

$$\frac{\text{ginger ale}}{\text{orange juice concentrate}} = \frac{32}{12} = \frac{g}{30}$$

(1st ratio, 2nd ratio)

$$\frac{g}{30} = \frac{32}{12}$$

$$30 \cdot \frac{g}{30} = 30 \cdot \frac{32}{12}$$

$$g = \frac{30 \cdot 32}{12} = \frac{960}{12}$$

$$g = 80$$

You should use 80 ounces of ginger ale for 30 ounces of frozen orange juice concentrate.

Solve.

7. Pecans cost $8.25 for 3 pounds. What is the cost of 5 pounds of pecans?

   $$\frac{\text{dollars}}{\text{pounds}} = \underline{} = \frac{c}{}$$

   $c = $ _____

8. Mandy drove 90 miles in 2 hours at a constant speed. How long would it take her to drive 225 miles at the same speed?

   $$\frac{\text{miles}}{\text{hours}} = \underline{} = \frac{}{h}$$

   $h = $ _____

9. Last week Geraldo bought 7 pounds of apples for $5.95. This week apples are the same price, and he buys 4 pounds. How much will he pay?

   _____

10. Aretha can type 55 words per minute. At that rate, how long will it take her to type a letter containing 935 words?

    _____

Name _____ Date _____ Class _____

LESSON 5-5 Challenge
Directly Speaking

Two quantities vary directly if they change in the same direction.
- If one quantity increases, then the other increases.
- If one quantity decreases, then the other decreases.

Here are some examples of direct variation.
- Amount bought and total cost
- Number of inches on a map and number of actual miles

If Farmer Jones plants 8 acres that produce 144 crates of melons, how many acres will produce 1,152 crates of melons?

Use the following proportion to solve.

$$\frac{\text{acres for 144 crates}}{\text{144 crates}} = \frac{\text{acres for 1,152 crates}}{\text{1,152 crates}}$$

$$\frac{8}{144} = \frac{x}{1,152} \qquad 144x = 9,216 \qquad x = 64$$

It will take 64 acres to produce 1,152 crates of melons.

Complete the proportion to find the answer.

		Do quantities increase or decrease?	Proportion	Answer
1.	If a furnace uses 40 gallons of oil in 8 days, how many gallons does it use in 10 days?		$\frac{\text{gal}}{8\text{ d}} = \frac{x\text{ gal}}{\text{d}}$	
2.	If 20 yards of wire weigh 80 pounds, what is the weight of 2 yards of the same wire?		$\frac{20\text{ yd}}{\text{lb}} = \frac{}{x\text{ lb}}$	
3.	A cookie recipe calls for 2.5 cups of sugar and 4 eggs. If 6 eggs are used, how much sugar is needed?		$\frac{c}{4} = \frac{c}{}$	
4.	The scale on a map is 1 inch = 60 miles. What distance on the map represents 300 miles?		$\frac{}{} = \frac{x\text{ in.}}{300\text{ mi}}$	
5.	At the market, 4 limes cost $1.50. How much will 10 limes cost?		$\frac{}{} = \frac{10}{}$	

Holt Mathematics

LESSON 5-5

Problem Solving
Solving Proportions

Write the correct answer.

1. Euros are currency used in several European countries. On one day in October 2005, you could exchange $3 for about 2.5 euros. How many dollars would you have needed to get 8 Euros?

2. A 3-ounce serving of tuna fish provides 24 grams of protein. How many grams of protein are in a 10-ounce serving of tuna fish?

3. Hooke's law states that the distance a spring is stretched is directly proportional to the force applied. If 20 pounds of force stretches a spring 4 inches, how much will the spring stretch if 80 pounds of force is applied?

4. Beeswax used in making candles is produced by honeybees. The honeybees produce 7 pounds of honey for each pound of wax they produce. How many pounds of honey is produced if 145 pounds of beeswax?

Choose the letter for the best answer.

5. For every 5 books her students read, Mrs. Fenway gives them a free homework pass for 4 days. Juan has accumulated homework passes for 12 days so far. What proportion would you write to find how many books Juan has read?

   A $\frac{4}{12} = \frac{x}{5}$

   B $\frac{4}{5} = \frac{x}{12}$

   C $\frac{4}{5} = \frac{12}{x}$

   D $\frac{5}{12} = \frac{4}{x}$

6. In his last 13 times at bat in the township baseball league, Santiago got 8 hits. If he is at bat 65 times for the season, how many hits will he get if his average stays the same?

   F $\frac{8}{65} = \frac{x}{13}$

   G $\frac{x}{65} = \frac{13}{8}$

   H $\frac{8}{x} = \frac{65}{13}$

   J $\frac{8}{13} = \frac{x}{65}$

7. A 12-pack of 8-ounce juice boxes costs $5.40. How much would an 18-pack of juice boxes cost if it is proportionate in price?

   A $9.40
   B $8.10
   C $3.60
   D $12.15

8. Jeanette can swim 105 meters in 70 seconds. How far can she probably swim in 30 seconds?

   F 20 meters
   G 245 meters
   H 45 meters
   J 55 meters

Reading Strategies
5-5 Draw a Conclusion

There is a quick method to check to see if two ratios are equal. It is called the **cross products** method. Follow these steps.

Step 1: Multiply factors that cross diagonally in the two ratios.

$\frac{4}{5} \times \frac{12}{15}$ ← 5×12
← 4×15

Step 2: If the products of cross factors are the same, the two ratios are equal.

The cross products of 60 are the same.

So, $\frac{4}{5}$ and $\frac{12}{15}$ are equal ratios. → $\frac{4}{5} \times \frac{12}{15}$ $\genfrac{}{}{0pt}{}{60}{60}$

Two equal ratios form a **proportion**. $\frac{4}{5} = \frac{12}{15}$ is a proportion.

If cross products are not the same, the two ratios are not equal.

Is $\frac{4}{6} \stackrel{?}{=} \frac{5}{9}$?

$\frac{4}{6} \times \frac{5}{9}$ $5 \times 6 = 30$
$4 \times 9 = 36$

The cross products are not equal.

$\frac{4}{6} = \frac{5}{9}$ **is not** a proportion, so $\frac{4}{6} \neq \frac{5}{9}$.

Use the cross products method to tell whether the two ratios are equal. Show your work. Write yes or no.

1. $\frac{2}{4}$ and $\frac{6}{12}$ _____

2. $\frac{2}{5}$ and $\frac{5}{10}$ _____

3. $\frac{6}{9}$ and $\frac{8}{12}$ _____

4. $\frac{3}{9}$ and $\frac{4}{10}$ _____

5. How can you tell if two ratios form a proportion?

_____

Puzzles, Twisters & Teasers
LESSON 5-5: The Proper Proportions!

Decide whether each pair of ratios is a proportion. Circle the letter above your answer. Then start with number 1, and use the letters to solve the riddle.

1. $\frac{2}{5} = \frac{6}{15}$
   - A correct
   - J incorrect

2. $\frac{6}{10} = \frac{36}{60}$
   - L correct
   - P incorrect

3. $\frac{4}{7} = \frac{5}{6}$
   - K correct
   - L incorrect

4. $\frac{4}{8} = \frac{12}{24}$
   - T correct
   - M incorrect

5. $\frac{1}{3} = \frac{86}{255}$
   - N correct
   - H incorrect

6. $\frac{39}{4} = \frac{121}{12}$
   - B correct
   - E incorrect

7. $\frac{2}{15} = \frac{12}{90}$
   - F correct
   - V incorrect

8. $\frac{18}{90} = \frac{1}{5}$
   - A correct
   - C incorrect

9. $\frac{45}{9} = \frac{15}{3}$
   - N correct
   - Z incorrect

10. $\frac{34}{6} = \frac{96}{16}$
    - L correct
    - S incorrect

11. $\frac{3}{24} = \frac{4}{52}$
    - K correct
    - L incorrect

12. $\frac{14}{20} = \frac{5}{8}$
    - G correct
    - E incorrect

13. $\frac{2}{5} = \frac{3}{12}$
    - U correct
    - F incorrect

14. $\frac{35}{4} = \frac{175}{20}$
    - T correct
    - E incorrect

Why did it get hot after the baseball game?

___ ___ ___   ___ ___ ___   ___ ___ ___ ___   ___ ___ ___ ___.

Name _____ Date _____ Class _____

Practice A
LESSON 5-6 Customary Measurements

Choose the letter of the best unit for each measurement.

1. The height of a house
   - A inches
   - B pounds
   - C feet
   - D miles

2. The weight of a letter
   - F ounces
   - G pounds
   - H inches
   - J tons

3. The capacity of a bathtub
   - A gallons
   - B cups
   - C fluid ounces
   - D pounds

4. The weight of a bowling ball
   - F ounces
   - G fluid ounces
   - H tons
   - J pounds

5. The thickness of a wallet
   - A ounces
   - B inches
   - C feet
   - D miles

6. The capacity of a spoon
   - F ounces
   - G fluid ounces
   - H cups
   - J gallons

Choose the letter of the best answer.

7. 8 gallons = ▉ quarts
   - A 2
   - B 4
   - C 16
   - D 32

8. 96 ounces = ▉ pounds
   - F 1536
   - G 16
   - H 8
   - J 6

9. 9 yards = ▉ feet
   - A 3
   - B 12
   - C 27
   - D 108

10. 4.5 cups = ▉ fluid ounces
    - F 2.25
    - G 9
    - H 36
    - J 72

11. 3 miles = ▉ feet
    - A 15,840
    - B 10,560
    - C 1584
    - D 36

12. 16,000 pounds = ▉ tons
    - F 4
    - G 8
    - H 80
    - J 1000

13. Awilda has 3 ft of ribbon.

   a. Write this measurement in inches.

   _____

   b. Suppose Awilda uses 20 in. of her ribbon to wrap a box. How much ribbon does she have left?

   _____ in., or _____ ft _____ in.

Name _____ Date _____ Class _____

LESSON 5-6 Practice B
Customary Measurements

Choose the most appropriate customary unit for each measurement. Justify your answer.

1. the weight of a paperback book

   _____

   _____

2. the capacity of a large soup pot

   _____

   _____

3. the length of a dining room table

   _____

   _____

4. the weight of an elephant

   _____

   _____

Convert each measure.

5. 6 mi to feet

6. 104 oz to pounds

7. 12 qt to pints

8. 5,000 lb to tons

9. 48 yd to feet

10. 96 fl oz to pints

11. 6.5 ft to inches

12. 20 qt to gallons

13. $3\frac{1}{4}$ lb to ounces

14. Marina has 2.5 lb of cashews. She puts 6 oz of cashews in a bag and gives the bag to her brother. What weight of cashews does Marina have left?

15. Faye is 5 ft 5 in. Faye is 10 in. shorter than her older brother. How tall is Faye's older brother?

Name _____ Date _____ Class _____

LESSON 5-6 Practice C
Customary Measurements

Choose the most appropriate customary unit for each measurement. Justify your answer.

1. the weight of an encyclopedia

2. the distance between two cities

3. the capacity of a medicine cup

4. the length of a hamster

Convert each measure.

5. 4.6 tons to pounds

6. 6600 ft to miles

7. 21 qt to pints

8. 5 yd to inches

9. 40 pt to gallons

10. 148 oz to pounds

Compare. Write <, >, or = .

11. 25,000 ft ☐ 5 mi   12. 6 lb ☐ 96 oz   13. 7.5 c ☐ 56 fl oz
14. 4 ft ☐ 32 in.   15. 2 gal ☐ 48 c   16. 4 yd ☐ 136 in.

17. Evan has 1 gal of orange juice. He uses 30 fl oz of orange juice to make a batch of smoothies. How much orange juice does Evan have left?

18. Helen has 3.5 yd of fabric. She buys 10 additional feet of fabric. How much fabric does Helen have now?

Name _____ Date _____ Class _____

LESSON 5-6

Reteach
Customary Measurements

You can use the facts in this table to help you convert customary units.

Length	Weight	Capacity
12 in. = 1 ft	16 oz = 1 lb	8 fl oz = 1 c
3 ft. = 1 yd	2,000 lb = 1 ton	2 c = 1 pt
5,280 ft = 1 mi		2 pt = 1 qt
		4 qt = 1 gal

To change larger units to smaller units, multiply.
3 lb = ☐ oz
3 × 16 = 48
3 lb = 48 oz

> 1 lb = 16 oz
> Multiply the number of pounds by 16.

To change smaller units to larger units, divide.
60 in. = ☐ ft
60 ÷ 12 = 5
60 in. = 5 ft

> 12 in. = 1 ft
> Divide the number of inches by 12.

Convert each measure.

1. 12 qt = ☐ gal

   Smaller unit → larger unit

   Fact: _____ qt = 1 gal

   Operation: _____

   12 ÷ 4 = _____

   12 qt = _____ gal

2. 9 yd = ☐ ft

   Larger unit → smaller unit

   Fact: 1 yd = _____ ft

   Operation: _____

   9 × _____ = _____

   9 yd = _____ ft

3. 7 lb = ☐ oz

   _____ _____ unit → _____ _____ unit

   7 lb = _____ oz

4. 48 fl oz = ☐ c

   _____ _____ unit → _____ _____ unit

   48 fl oz = _____ c

5. 8 ft = _____ in.

6. 4,000 lb = _____ tons

48

Holt Mathematics

Name _____ Date _____ Class _____

Challenge
LESSON 5-6 *Conversion Diversion*

The table shows data about some of the most famous baseball parks in the United States. Use the table to answer the questions.

Baseball Field	Team	Year Opened	Year Closed	Cost	Dimensions of the Center Field Line
Ebbets Field	Brooklyn Dodgers	1913	1957	$750,000	393 ft
Forbes Field	Pittsburgh Pirates	1909	1970	$1 million	400 ft
Tiger Stadium	Detroit Tigers	1912	1999	$8 million	440 ft
Yankee Stadium	New York Yankees	1923		$50 million	408 ft
L.A. Coliseum	Los Angeles Dodgers	1958	1961	$950,000	420 ft

1. How many inches was the center field line in Tiger Stadium?

2. How many thousands of dollars did it cost to build Yankee Stadium?

3. For about how many months was the L.A. Coliseum open?

4. How many quarters (25¢) would it have taken to build the L.A. Coliseum?

5. How many inches longer was center field in Forbes Field than in Ebbets Field?

6. How many miles to the nearest hundredth of a mile was the center field line in the L.A. Coliseum?

7. For about how many months was Forbes Field open?

8. How many yards is the center field line in Yankee Stadium?

Name _____ Date _____ Class _____

LESSON 5-6 Problem Solving
Customary Measurements

Write the correct answer.

1. In 2003, a popcorn sculpture of King Kong was displayed in London. The sculpture was 13 ft tall and 8.75 ft wide. How many inches wide was the sculpture?

2. A pilot whale weighs 1500 lb. A walrus weighs 1.6 tons. Which weighs more? How much more?

3. A zoo has a rhesus monkey that weighed 20 lb. The monkey became sick and lost 18 oz. What was the monkey's new weight?

4. Two containers have capacities of 192 fl oz and 1.25 gal. Which container has a greater capacity? How much greater?

Choose the letter for the best answer.

This table gives lengths and weights for some apes.

Apes		
Name	Maximum Height	Maximum Weight
Chimpanzee	4 ft	115 lb
Gorilla	67.2 in.	0.2 tons
Orangutan	1.5 yd	3200 oz
Siamang	36 in.	240 oz

5. Which ape has the greatest weight?
   A Chimpanzee
   B Gorilla
   C Orangutan
   D Siamang

6. Which ape has the least height?
   F Chimpanzee
   G Gorilla
   H Orangutan
   J Siamang

7. Which ape has a maximum weight of 200 lb?
   A Chimpanzee
   B Gorilla
   C Orangutan
   D Siamang

8. Which two apes have a 6-inch difference in height?
   F Chimpanzee and gorilla
   G Gorilla and orangutan
   H Siamang and chimpanzee
   J Orangutan and chimpanzee

Name _____ Date _____ Class _____

LESSON 5-6 Reading Strategies
Use a Flowchart

This flowchart shows two ways to solve this problem: 56 oz = ■ pounds.

```
Identify equivalent measures that have the
same units as the measures in the problem.
            16 oz = 1 lb
                ↓
     Choose a method to convert units.
         ↙                    ↘
```

Write a proportion.
Use a ratio of equivalent measures and a ratio of the measures in the problem.

ounces → $\dfrac{16}{1} = \dfrac{56}{x}$
pounds →

Write a multiplication equation.
Use a ratio of equivalent measures, which equals 1. Set up the ratio so that you can cancel units.

$56 \text{ oz} = 56 \text{ oz} \times \dfrac{1 \text{ lb}}{16}$

Solve to find the value of x.
$16 \cdot x = 56 \cdot 1$
$16x = 56$
$x = 3.5$

Simplify. $56 \text{ oz} = \dfrac{56}{1} \times \dfrac{1 \text{ lb}}{16}$
$= \dfrac{56 \text{ lb}}{16}$
$= 3.5 \text{ lb}$

56 oz = 3.5 pounds

Use the problem 18 yd = ■ ft for Exercises 1–6.

1. What equivalent measures can you use to solve this problem? _____

2. What proportion can you write to solve this problem? _____

3. Solve the proportion you wrote. _____

4. What multiplication equation can you write to solve this problem?

_____

5. Simplify the equation you wrote. _____

6. How do your answers to Exercises 3 and 5 compare?

_____

Puzzles, Twisters, & Teasers
LESSON 5-6 — Because It's Customary

Choose the most appropriate customary unit for each measurement. Circle the letter above your answer.

1. the capacity of a hot water heater
   - B fluid ounces
   - **T gallons**

2. the weight of a bowling ball
   - **D pounds**
   - K tons

3. the thickness of a book
   - **N inches**
   - C feet

4. the capacity of a thermos
   - A gallons
   - **H quarts**

Next, choose the measurement that is equivalent to the one given. Circle the letter above your answer.

5. 6 ft = ?
   - S 48 in.
   - **W 72 in.**

6. 12 oz = ?
   - **C 0.75 lb**
   - O 192 lb

7. 80 fl oz = ?
   - **E 10 c**
   - C 5 lb

8. 3.5 tons = ?
   - P 56 lb
   - **O 7000 lb**

9. 36 ft = ?
   - **K 12 yd**
   - O 3 yd

10. 40 qt = ?
    - I 160 gal
    - **Y 10 gal**

Write the circled letters above the problem numbers to solve the riddle.

Why do lions eat raw meat?

T   H   E   Y   D   O   N   T   K   N   O   W
1. 4. 7. 10. 2. 8. 3. 1. 9. 3. 8. 5.

H   O   W   T   O   C   O   O   K
4. 8. 5. 1. 8. 6. 8. 8. 9.

Name _____ Date _____ Class _____

Practice A
LESSON 5-7 Similar Figures and Proportions

Identify the corresponding sides.

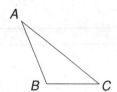

1. AB corresponds to _____.

2. BC corresponds to _____.

3. AC corresponds to _____.

Identify the corresponding sides. Then use ratios to determine whether the triangles are similar.

4.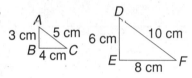

_____

_____

5.
J 5 in. /\ 5 in.   M 6 in. /\ 6 in.
K ―――― L   N ―――― P
    7 in.   8 in.

_____

_____

6.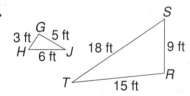

_____

_____

7.
    Y   U
4 m /\ 4 m   5 m /\ 5 m
X ――――   V ―――― W
    4 m Z   5 m

_____

_____

Use the properties of similarity to determine whether the figures are similar.

8.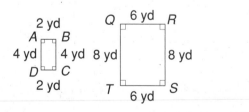

9.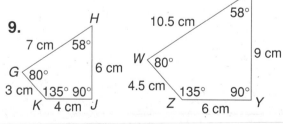

_____

_____

Copyright © by Holt, Rinehart and Winston.
All rights reserved.

Holt Mathematics

Name _____ Date _____ Class _____

LESSON 5-7 Practice B
Similar Figures and Proportions

Identify the corresponding sides in each pair of triangles.
Then use ratios to determine whether the triangles are similar.

1.
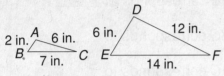

_____

_____

2.

_____

_____

3.
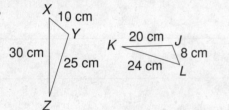

_____

_____

4.

_____

_____

Use the properties of similarity to determine whether the figures are similar.

5.
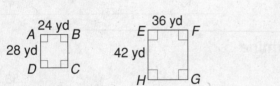

_____

_____

6.

_____

_____

Copyright © by Holt, Rinehart and Winston.
All rights reserved.

Holt Mathematics

Name _____ Date _____ Class _____

LESSON 5-7 Practice C
Similar Figures and Proportions

Use the properties of similarity to determine whether the figures are similar.

1. X, 3.2 m, 4.8 m, Y, 6.4 m, Z; R, 1.6 m, 2.4 m, S, 3.2 m, T

2. A, 2.2 ft, 2.5 ft, B, 1.7 ft, C; D, 3.4 ft, E, 5.5 ft, 5.0 ft, F

_____

_____

3. W, 12 yd, X, 10 yd, 10 yd, Z, 6 yd, Y; M, 15 yd, N, 8 yd, 16 yd, P, 15 yd, O

4. E, 14 cm, F, 104°, 76°, 14 cm, 14 cm, H, 76°, 104°, G, 14 cm; J, 21 cm, K, 108°, 72°, 21 cm, 21 cm, M, 72°, 108°, L, 21 cm

_____

_____

The figures in each pair are similar. Find the missing lengths or angle measures.

5. A, 14.4 in., 15.6 in., B, ?, C; X, 12 in., Y, 5 in., Z, 13 in.

6. D, 15 m, 29°, E, 103°, 10 m, 48°, 20 m, F, 26 m, M, ?, 19.5 m, ?, N, 13 m, L, ?

_____

_____

7.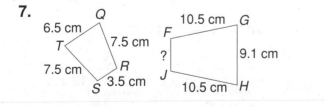

8.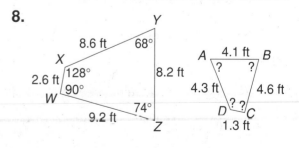

_____

_____

55   Holt Mathematics

LESSON 5-7 Reteach
Similar Figures and Proportions

Figures that have the same shape but not the same size are **similar figures**. In similar figures, the ratio of the lengths of the corresponding sides are proportional, and the corresponding angles have equal measures.

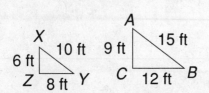

To determine if △ABC is similar to △XYZ, you can write a proportion for each pair of corresponding sides.

longest sides	middle sides	shortest sides
$\dfrac{AB}{XY} = \dfrac{15}{10} = \dfrac{3}{2}$	$\dfrac{BC}{YZ} = \dfrac{12}{8} = \dfrac{3}{2}$	$\dfrac{AC}{XZ} = \dfrac{9}{6} = \dfrac{3}{2}$

The corresponding sides are always in the ratio $\dfrac{3}{2}$. So the triangles are similar.

If a polygon has more than 3 sides, you must also show that the corresponding angles are equal.

Identify the corresponding sides. Use ratios to determine whether the figures are similar.

1.

$\dfrac{TU}{EF} = \dfrac{}{8} = \dfrac{}{1}$; $\dfrac{SU}{} = \dfrac{}{} = \dfrac{}{}$;

$\dfrac{ST}{} = \dfrac{}{} = \dfrac{}{}$

Are the ratios proportional? _____

Are the triangles similar? _____

2.

$\dfrac{PQ}{} = \dfrac{}{} = \dfrac{}{}$; $\dfrac{PR}{} = \dfrac{}{} = \dfrac{}{}$;

$\dfrac{QR}{} = \dfrac{}{} = \dfrac{}{}$

Are the ratios proportional? _____

Are the triangles similar? _____

3.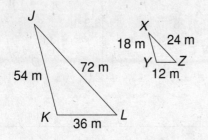

4.

Holt Mathematics

Name _____ Date _____ Class _____

Challenge
LESSON 5-7 The Same, Only Bigger

You can sometimes create a similar figure by using copies of the original figure.

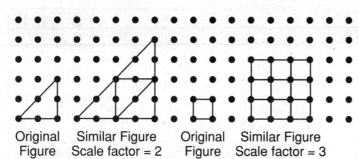

Notice that the scale factor tells you how many times to repeat the original figure along each side or edge of the similar figure.

Use the given scale factor and copies of the original figure to draw a figure similar to the original figure.

1. scale factor = 4

2. scale factor = 2

3.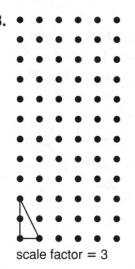
scale factor = 3

4.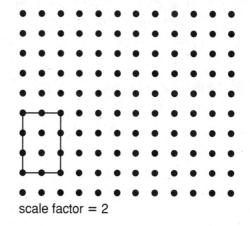
scale factor = 2

5. Draw a figure in the space below. Use a scale factor of 2 to create a similar figure.

Name _____ Date _____ Class _____

LESSON 5-7 Problem Solving
Similar Figures and Proportions

Use the information in the table to solve problems 1–3.

1. A small reproduction of one of the paintings in the list is similar in size. The reproduction measures 11 inches by 10 inches. Of which painting is this a reproduction?

   _____

Painting	Artist	Original Size (in.)
Mona Lisa	Leonardo da Vinci	30 by 21
The Dance Class	Edgar Degas	33 by 30
The Blue Vase	Paul Cézanne	22 by 18

2. A local artist painted a reproduction of Cézanne's painting. It measures 88 inches by 72 inches. Is the reproduction similar to the original? What is the ratio of corresponding sides?

   _____

3. A poster company made a poster of da Vinci's painting. The poster is 5 feet long and 3.5 feet wide. Is the poster similar to the original Mona Lisa? What is the ratio of corresponding sides?

   _____

Choose the letter for the best answer.

4. Triangle ABC has sides of 15 cm, 20 cm, and 25 cm. Which triangle could be similar to triangle ABC?

   A A triangle with sides of 3 cm, 4 cm, and 5 cm
   B A triangle with sides of 5 cm, 6 cm, and 8 cm
   C A triangle with sides of 30 cm, 40 cm, and 55 cm
   D A triangle with sides of 5 cm, 10 cm, and 30 cm

5. A rectangular picture frame is 14 inches long and 4 inches wide. Which dimensions could a similar picture frame have?

   F Length = 21 in.; width = 8 in.
   G Length = 35 in.; width = 15 in.
   H Length = 49 in.; width = 14 in.
   J Length = 7 in.; width = 3 in.

6. A rectangle is 12 meters long and 21 meters wide. Which dimensions correspond to a nonsimilar rectangle?

   A 4 m; 7 m   C 20 m; 35 m
   B 8 m; 14 m   D 24 m; 35 m

7. A rectangle is 6 feet long and 15 feet wide. Which dimensions correspond to a similar rectangle?

   F 8 ft; 24 ft   H 15 ft; 35 ft
   G 10 ft; 25 ft   J 18 ft; 40 ft

Name _____ Date _____ Class _____

LESSON 5-7 Reading Strategies
Understanding Vocabulary

Similar means almost the same. If two objects are similar, they have some things in common.

Similar figures are figures that are nearly the same. Similar figures have the same shape, but are different sizes.

Similar figures have **corresponding sides** and **corresponding angles**. *Corresponding* means matching. Each side and angle in a similar figure has a corresponding side and angle.

These two triangles are similar.

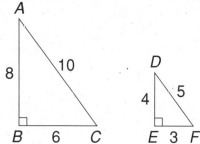

Use the figures above to answer each question.

1. What angle corresponds to angle *B*?

   _____

2. What angle corresponds to angle *A*?

   _____

3. What side corresponds to side *BC*?

   _____

Are the figures similar? Answer yes or no for each pair.

4. _____

5. _____

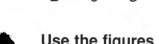

Copyright © by Holt, Rinehart and Winston.
All rights reserved.

59

Holt Mathematics

Name _____ Date _____ Class _____

LESSON 5-7 Puzzles, Twisters & Teasers
Concentrating on Figures

Pretend this is a game of concentration. The object of the game is to match cards with similar figures. Each box represents a card with a figure on it. When you match 2 cards, cross them out. Rearrange the letters of the unmatched cards to solve the riddle.

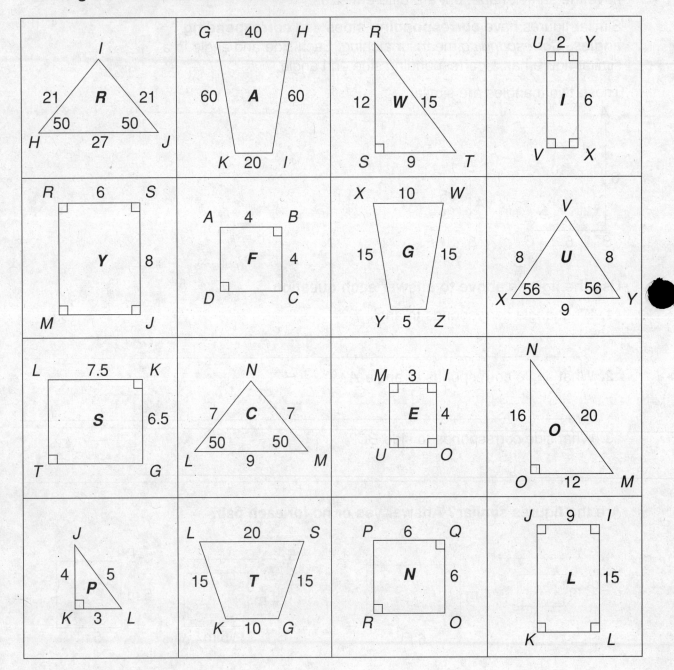

What kind of flowers are on your face?

___ ___ ___ ___ ___ ___

Name _____ Date _____ Class _____

LESSON 5-8 Practice A
Using Similar Figures

For each pair of similar figures write a proportion containing the unknown length. Then solve.

1.
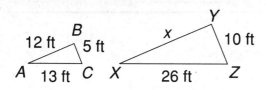

_____

2.
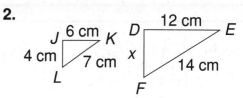

_____

3.
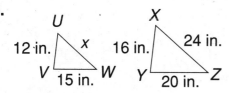

_____

4.
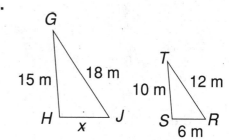

_____

5. Kareem and Julio have rectangular model train layouts that are similar to each other. Julio's layout is 4 feet by 7 feet. Kareem's layout is 6 feet wide. What is the length of Kareem's layout?

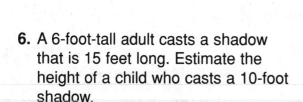

_____

6. A 6-foot-tall adult casts a shadow that is 15 feet long. Estimate the height of a child who casts a 10-foot shadow.

_____

Name _____ Date _____ Class _____

LESSON 5-8 Practice B
Using Similar Figures

△ABC ~ △DEF in each pair. Find the unknown lengths.

1.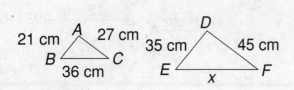

2.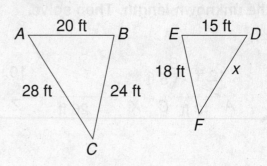

3.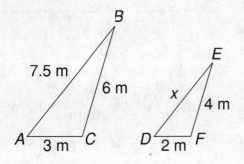

4.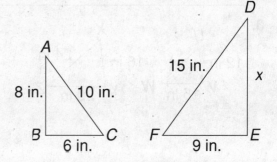

5. The two rectangular picture frames at the right are similar. What is the height of the larger picture frame?

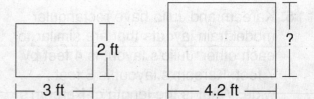

6. A palm tree casts a shadow that is 44 feet long. A 6-foot ladder casts a shadow that is 16 feet long. Use Estimate the height of the palm tree.

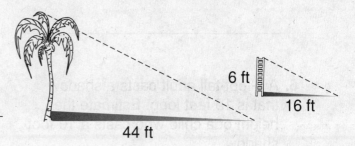

Holt Mathematics

Name _____ Date _____ Class _____

LESSON 5-8 Practice C
Using Similar Figures

Find the unknown length in each pair of similar figures.

1.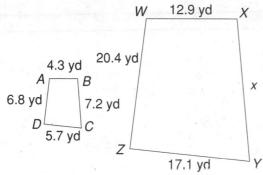

2.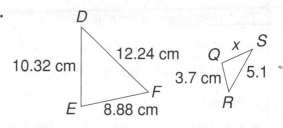

3.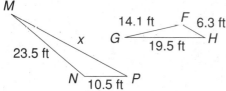

4.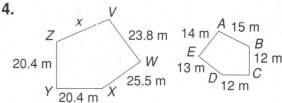

Estimate the height of each object in the picture below.

5. house _____

6. tree _____

7. lamppost _____

8. radio tower _____

Holt Mathematics

LESSON 5-8 Reteach
Using Similar Figures

If you know that 2 figures are similar, you can use proportions to find unknown lengths of sides.

The triangles are similar.

   Side *AC* corresponds to side *DF*.

   Side *AB* corresponds to side *DE*.

   Side *BC* corresponds to side *EF*.

Write a proportion comparing the lengths of a pair of corresponding sides.

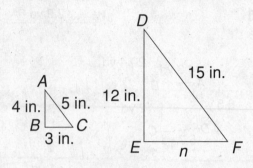

$$\frac{AC}{DF} = \frac{BC}{EF}$$

$$\frac{5}{15} = \frac{3}{n}$$

$$5 \cdot n = 15 \cdot 3$$

$$5n = 45$$

$$\frac{5n}{5} = \frac{45}{5}$$

$$n = 9$$

The length of the missing side is 9 in.

Find the unknown length in each pair of similar figures.

1.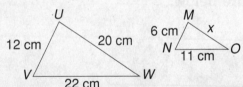

   $\dfrac{UW}{} = \dfrac{UV}{}$; $\dfrac{20}{} = \dfrac{12}{}$

   x = _____

2.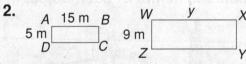

   $\dfrac{WZ}{} = \dfrac{WX}{}$; $\dfrac{9}{} = \dfrac{}{}$

   y = _____

3.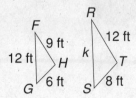

   k = _____

4.

   s = _____

Name _____ Date _____ Class _____

LESSON 5-8 Challenge
You Be the Artist

Your club is planning a poster to advertise the school's international dinner. The poster will be enlarged and used as a mural on the school cafeteria wall. The poster will also be reduced and used as flyers. The mural will be 10 feet high and 15 feet long. The flyers will be printed on $8\frac{1}{2}$-by-11-inch paper.

Plan the size of the poster so that the enlargement and reduction will be easy to make.

1. What is the width-to-length ratio for the wall mural? Write the ratio in simplest terms.

   _____

2. What is the width-to-length ratio for the flyer? Write the ratio in simplest terms.

   _____

3. Do you want your artwork to fill the entire page for the flyer?

   _____

4. What are some possible dimensions for your poster?

   _____

5. Will your poster fill the wall space when it is enlarged for the mural? Explain.

   _____

6. What would be a good size for the artwork on the flyer?

   _____

LESSON 5-8 Problem Solving
Using Similar Figures

Write the correct answer.

1. An architect is building a model of a tennis court for a new client. On the model, the court is 6 inches wide and 13 inches long. An official tennis court is 36 feet wide. What is the length of a tennis court?

2. Mr. Hemley stands next to the Illinois Centennial Monument at Logan Square in Chicago and casts a shadow that is 18 feet long. The shadow of the monument is 204 feet long. If Mr. Hemley is 6 feet tall, how tall is the monument?

3. The official size of a basketball court in the NBA is 94 feet by 50 feet. The basketball court in the school gym is 47 feet long. How wide must it be to be similar to an NBA court?

4. Two rectangular desks are similar. The larger one is 42 inches long and 18 inches wide. The smaller one is 35 inches long. What is the width of the smaller desk?

Choose the letter for the best answer.

5. An isosceles triangle has two sides that are equal in length. Isosceles triangle ABC is similar to isosceles triangle XYZ. What proportion would you use to find the length of the third side of triangle XYZ?

   A $\frac{BC}{XZ} = \frac{AB}{XY}$ C $\frac{AB}{XY} = \frac{AC}{XZ}$

   B $\frac{AC}{XY} = \frac{BC}{XZ}$ D $\frac{AB}{XY} = \frac{BC}{YZ}$

6. The dining room at Monticello, Thomas Jefferson's home in Virginia, is 216 inches by 222 inches. Of the following, which size rug would be similar in shape to the dining room?

   F 72 inches by 74 inches
   G 108 inches by 110 inches
   H 118 inches by 111 inches
   J 84 inches by 96 inches

7. A 9-foot street sign casts a 12-foot shadow. The lamppost next to it casts a 24-foot shadow. How tall is the lamppost?

   A 24 feet
   B 15 feet
   C 18 feet
   D 36 feet

Reading Strategies
5-8 Use a Flowchart

It is very difficult to measure the height of tall tree or a utility pole directly. You can set up proportions to measure very tall objects indirectly.

This method of measuring is called **indirect measurement**. You do not actually measure the height. You use a proportion to find the height.

These two triangles are similar.
Find length x in triangle DEF.

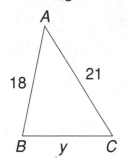
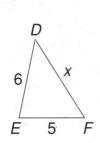

The flowchart helps you set up a proportion to find the value of x.

Write a proportion to find the missing length. $\frac{AB}{DE} = \frac{AC}{DF}$
↓
Put the lengths of the known sides into the equation. x stands for the length of the unknown side. $\frac{18}{6} = \frac{21}{x}$
↓
Multiply cross products. $18 \cdot x = 6 \cdot 21$
↓
Find the value of x. $\frac{18x}{18} = \frac{6 \cdot 21}{18}$ $x = \frac{126}{18}$ $x = 7$

Answer each question.

1. Why is this method is called indirect measurement?

   _____
   _____

2. What is the next step after setting up the proportion?

   _____

3. Write a proportion to find the length y in triangle ABC.

   _____

Copyright © by Holt, Rinehart and Winston.
All rights reserved.

Holt Mathematics

Name _____ Date _____ Class _____

LESSON 5-8 Puzzles, Twisters & Teasers
Measure This!

Use indirect measurement to find the width of the smaller rectangle. Show the steps you used by filling in the blanks in the sentences. To solve the riddle, find the letter(s) in each answer with a number below it. Match the letters to the numbered blanks in the riddle.

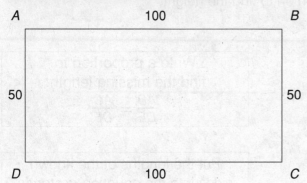

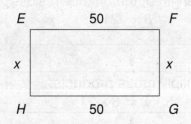

1. Write a __ __ __ __ __ __ __ __ __ __.
                    2

2. Substitute the __ __ __ __ __ __ of the sides.
                           6

3. Find the __ __ __ __ __   __ __ __ __ __ __ __ __.
               4   5   3

4. __ __ __ __ __ for x.
    1

5. Divide.

6. $x = $ ____

What do you call a cat that swims and has eight legs?

An ___ ___ ___ ___-___ ___ ___ ___
    2 4 6 2   5 3 1 1

Name _____ Date _____ Class _____

Practice A
LESSON 5-9 Scale Drawings and Scale Models

Identify the scale factor. Choose the best answer.

1. Person: 72 inches
   Action figure: 6 inches

   A $\frac{1}{7}$ C $\frac{1}{12}$
   B $\frac{1}{10}$ D $\frac{1}{15}$

2. Dog: 24 inches
   Stuffed animal: 8 inches

   F $\frac{1}{3}$ H $\frac{1}{5}$
   G $\frac{1}{4}$ J $\frac{1}{6}$

3. Fish: 16 inches
   Fishing lure: 2 inches

   A $\frac{1}{6}$ C $\frac{1}{12}$
   B $\frac{1}{8}$ D $\frac{1}{14}$

4. House: 30 feet
   Dollhouse: 3 feet

   F $\frac{1}{3}$ H $\frac{1}{27}$
   G $\frac{1}{10}$ J $\frac{1}{33}$

Identify the scale factor.

5.

	Guitar	Ukulele
Length (in.)	36	18

6.

	Car	Toy Car
Length (ft)	12	3

7.

	Flute	Piccolo
Length (in.)	30	10

8.

	Poodle	Toy Poodle
Height (in.)	56	8

9. On a road map of New York, the distance from New York City to Albany is 3 inches. What is the actual distance between the cities if the map scale is 1 inch = 50 miles?

10. On a scale drawing, a bookshelf is 8 inches tall. The scale factor is $\frac{1}{8}$. What is the height of the bookshelf?

Practice B
Lesson 5-9: Scale Drawings and Scale Models

Identify the scale factor.

1.
	Alligator	Toy Alligator
Length (in.)	175	7

2.
	Airplane	Model
Length (ft)	24	3

3.
	Car	Toy Car
Length (ft)	13.5	1.5

4.
	Person	Action Figure
Height (in.)	66	6

5.
	Boat	Model
Length (in.)	128	8

6.
	Fish	Fishing Lure
Length (in.)	18	2

7.
	Tiger	Stuffed Animal
Length (in.)	70	14

8.
	House	Dollhouse
Height (ft)	39.2	2.8

9. On a scale drawing, a school is 1.6 feet tall. The scale factor is $\frac{1}{22}$. Find the height of the school.

10. On a road map of Pennsylvania, the distance from Philadelphia to Washington, D.C., is 6.8 centimeters. What is the actual distance between the cities if the map scale is 2 centimeters = 40 miles?

11. On a scale drawing, a bicycle is $6\frac{4}{5}$ inches tall. The scale factor is $\frac{1}{6}$. Find the height of the bicycle.

Name _____ Date _____ Class _____

Practice C
LESSON 5-9 Scale Drawings and Scale Models

Identify the scale factor.

1.
	Bear	Stuffed Animal
Height (in.)	62	15.5

2.
	House	Dollhouse
Height (ft)	32.4	2.7

3.
	Airplane	Model
Length (ft)	25.5	1.5

4.
	Alligator	Toy Alligator
Length (in.)	128.1	6.1

The scale factor of each model is 1:16. Find the missing dimensions.

	Item	Actual Dimensions	Model Dimensions
5.	barn	height: 32 ft length:	height: length: 3.5 ft
6.	submarine	length:	length: $18\frac{3}{4}$ ft
7.	bookcase	height: 96 in.	height:
8.	tree	height:	height: $2\frac{1}{2}$ ft
9.	car	length: 13 ft height: 5.5 ft	length: height:
10.	shark	length:	length: $14\frac{1}{4}$ in.

11. Hillary took a photograph of her house, which has an actual height of 28.5 feet. If the house measures 3.6 inches tall in the photograph, what is the scale factor? _____

12. On a road map, the distance from Portland to Seattle is 8 centimeters. What is the actual distance between the cities if the map scale is 2 centimeters = 37.5 miles? _____

13. A sculptor plans a statue by making a drawing to scale. On the drawing, the statue is $8\frac{2}{5}$ inches tall. The scale factor in the drawing is $\frac{1}{23}$. Find the height of the statue. _____

LESSON 5-9 Reteach
Scale Drawings and Scale Models

The dimensions of a scale model or scale drawing are related to the actual dimensions by a *scale factor*. The **scale factor** is a ratio.

The length of a model car is 9 in. ⟶ $\dfrac{9 \text{ in.}}{162 \text{ in.}} = \dfrac{9 \div 9}{162 \div 9} = \dfrac{1}{18}$
The length of the actual car is 162 in. ⟶

$\dfrac{9}{162}$ can be simplified to $\dfrac{1}{18}$. ⟵ The scale factor is $\dfrac{1}{18}$.

If you know the scale factor, you can use a proportion to find the dimensions of an actual object or of a scale model or drawing.

- The scale factor of a model train set is $\dfrac{1}{87}$. A piece of track in the model train set is 8 in. long. What is the actual length of the track?

$\dfrac{\text{model length}}{\text{actual length}} = \dfrac{8}{x} \qquad \dfrac{8}{x} = \dfrac{1}{87} \qquad x = 696$

The actual length of track is 696 inches.

- The distance between 2 cities on a map is 4.5 centimeters. The scale on the map is 1 cm = 40 miles.

$\dfrac{\text{distance on map}}{\text{actual distance}} = \dfrac{4.5 \text{ cm}}{x \text{ mi}} = \dfrac{1 \text{ cm}}{40 \text{ mi}} \qquad \dfrac{4.5}{x} = \dfrac{1}{40} \qquad x = 180$

The actual distance is 180 miles.

Identify the scale factor.

1. Photograph: height 3 in.
   Painting: height 24 in.

   $\dfrac{\text{photo height}}{\text{painting height}} = \dfrac{\text{in.}}{\text{in.}} = \underline{}$

2. Butterfly: wingspan 20 cm
   Silk butterfly: wingspan 4 cm

   $\dfrac{\text{silk butterfly}}{\text{butterfly}} = \dfrac{\text{cm}}{\text{cm}} = \underline{}$

3. On a scale drawing, the scale factor is $\dfrac{1}{12}$. A plum tree is 7 inches tall on the scale drawing. What is the actual height of the tree?

4. On a road map, the distance between 2 cities is 2.5 inches. The map scale is 1 inch = 30 miles. What is the actual distance between the cities?

Holt Mathematics

Name _____ Date _____ Class _____

LESSON 5-9 Challenge
Balls of Sports

Each circle below is a scale drawing of a different type of ball used in a sport.

- Measure the diameter of each circle to the nearest tenth of a centimeter.
- Use the scale to find the actual diameter of the ball to the nearest tenth of a centimeter.
- Use the chart below to find the sport in which the ball is used.

Diameter of Balls Used in Various Sports	
Basketball	24.0 cm
Baseball	7.5 cm
Golf	4.2 cm
Table Tennis	3.8 cm
Tennis	6.4 cm
Volleyball	21.0 cm

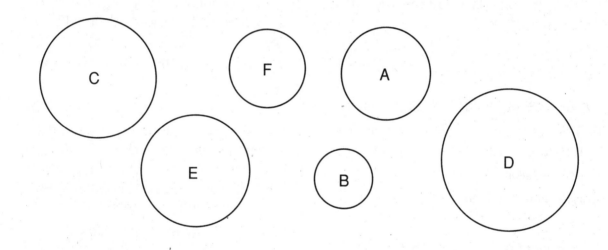

	Circle	Scale	Measured Diameter	Actual Diameter	Sport
1.	A	1 cm = 3 cm			
2.	B	1 cm = 15 cm			
3.	C	1 cm = 2 cm			
4.	D	1 cm = 1 cm			
5.	E	1 cm = 1.4 cm			
6.	F	1 cm = 10 cm			

Name _____ Date _____ Class _____

LESSON 5-9 Problem Solving
Scale Drawings and Scale Models

Write the correct answer.

1. The scale on a road map is 1 centimeter = 500 miles. If the distance on the map between New York City and Memphis is 2.2 centimeters, what is the actual distance between the two cities?

2. There are several different scales in model railroading. Trains designated as O gauge are built to a scale factor of 1:48. To the nearest hundredth of a foot, how long is a model of a 50-foot boxcar in O gauge?

3. For a school project, LeeAnn is making a model of the Empire State Building. She is using a scale of 1 centimeter = 8 feet. The Empire State Building is 1,252 feet tall. How tall is her model?

4. A model of the Eiffel Tower that was purchased in a gift shop is 29.55 inches tall. The actual height of the Eiffel Tower is 985 feet, or 11,820 inches. What scale factor was used to make the model?

Choose the letter for the best answer.

5. The scale factor for Maria's dollhouse furniture is 1:8. If the sofa in Maria's dollhouse is $7\frac{1}{2}$ inches long, how long is the actual sofa?
   A 54 inches
   B 60 inches
   C 84 inches
   D $15\frac{1}{2}$ inches

6. The Painted Desert is a section of high plateau extending 150 miles in northern Arizona. On a map, the length of this desert is 5 centimeters. What is the map scale?
   F 1 centimeter = 25 miles
   G 5 centimeters = 100 miles
   H 1 centimeter = 30 miles
   J 1 centimeter = 50 miles

7. Josh wants to add a model of a tree to his model railroad layout. How big should the model tree be if the actual tree is 315 inches and the scale factor is 1:90?
   A 395 inches
   B 39.5 inches
   C 35 inches
   D 3.5 inches

8. The scale on a wall map is 1 inch = 55 miles. What is the distance on the map between two cities that are 99 miles apart?
   F 44 inches
   G 1.8 inches
   H 2.5 inches
   J 0.55 inches

Name _____ Date _____ Class _____

LESSON 5-9 Reading Strategies
Read a Map

A **scale drawing** has the same shape, but is not the same size, as the object it represents. A map is an example of a scale drawing.

This is a map of a campground. The scale is 1 cm = 10 ft.

To find how far the campground entrance is from the canoe rental office, follow the steps. Use a centimeter ruler to measure.

campsite 1
water
campsite 2

ENTRANCE
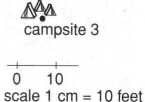
campsite 3
0 10
scale 1 cm = 10 feet

Step 1: Measure the distance in centimeters.
→ The distance is 4 centimeters.

Step 2: Set up a proportion using the map scale as one ratio.
→ $\frac{1 \text{ cm}}{10 \text{ ft}} = \frac{4 \text{ cm}}{x \text{ ft}}$

Step 3: Set up cross products. → $1x = 4 \cdot 10$

Step 4: Solve to find the value of x. → $x = 40$

Use the map to answer each question.

1. How many centimeters is Campsite 3 from the water?

   _____

2. Write a proportion to find the distance from Campsite 3 to the water.

   _____

3. How many centimeters is Campsite 3 from the canoe rental office?

   _____

4. Write a proportion to find the distance from Campsite 3 to the canoe rental office.

   _____

Name _____ Date _____ Class _____

LESSON 5-9
Puzzles, Twisters & Teasers
Let's Rock!

Find and circle these words in the word search. Find a word that solves the riddle. Circle it and write it on the line.

scale   model   drawing   factor   ratio
size   actual dimensions represent proportion

```
V T S P R O P O R T I O N
M N V S B R M K O L J I R
O Y D I M E N S I O N S E
D A X Z R A T I O F F C P
E O P E V O Q W E A A A R
L I K L P S C R T U C L E
A S D F G H M K Q R T E S
A C T U A L B V E Z O E E
N O D R A W I N G T R O N
A J D F A K L M N T S N T
```

What do you do to make a baby sleep on a space ship?

You ___ ___ ___ ___ ___ ___.

LESSON 5-1 Practice A: Ratios

Match the ratios.

A farmer has 5 pigs, 13 chickens, and 8 cows.
1. cows to pigs — 13:5
2. chickens to pigs — 5:8
3. cows to chickens — 8:5
4. pigs to cows — 8:13

The school orchestra has 9 cellos, 14 flutes, and 17 violins.
5. cellos to violins — 9 to 14
6. flutes to cellos — 9 to 17
7. violins to flutes — 17 to 14
8. cellos to violins — 14 to 9

Miguel has 8 pennies, 5 nickels, and 3 quarters.
9. nickels to pennies — $\frac{3}{8}$
10. pennies to quarters — $\frac{5}{3}$
11. nickels to quarters — $\frac{5}{3}$
12. quarters to pennies — $\frac{8}{3}$

A bowl has 16 grapes, 7 cherries, and 9 strawberries.
13. grapes to strawberries — 7:16
14. cherries to grapes — 7:9
15. strawberries to cherries — 16:9
16. cherries to strawberries — 9:7

A baseball team has 4 pitchers, 10 outfielders, and 12 infielders. Write each ratio in all three forms.

17. pitchers to infielders
$\frac{4}{12}$, 4 to 12, 4:12; or $\frac{1}{3}$, 1 to 3, 1:3

18. infielders to outfielders
$\frac{12}{10}$, 12 to 10, 12:10; or $\frac{6}{5}$, 6 to 5, 6:5

19. pitchers to outfielders
$\frac{4}{10}$, 4 to 10, 4:10; or $\frac{2}{5}$, 2 to 5; 2:5

20. outfielders to entire team
$\frac{10}{26}$, 10 to 26, 10:26 or $\frac{5}{13}$, 5 to 13; 5:13

21. Meg used 24 red tiles and 12 yellow tiles to make a design. Write the ratio of red tiles to yellow tiles in simplest form.
2 to 1

22. Tell which club has the greater ratio of girls to boys.
Art Club

	Movie Club	Art Club
Girls	16	8
Boys	14	4

LESSON 5-1 Practice B: Ratios

The annual dog show has 22 collies, 28 boxers, and 18 poodles. Write each ratio in all three forms.

1. collies to poodles
$\frac{22}{18}$, 22 to 18, 22:18; or $\frac{11}{9}$, 11 to 9, 11:9

2. boxers to collies
$\frac{28}{22}$, 28 to 22, 28:22; or $\frac{14}{11}$, 14 to 11, 14:11

3. poodles to boxers
$\frac{18}{28}$, 18 to 28, 18:28; or $\frac{9}{14}$, 9 to 14, 9:14

4. poodles to collies
$\frac{18}{22}$, 18 to 22, 18:22; or $\frac{9}{11}$, 9 to 11, 9:11

The Franklin School District has 15 art teachers, 27 math teachers, and 18 Spanish teachers. Write the given ratio in all three forms.

5. art teachers to math teachers
$\frac{15}{27}$, 15 to 27, 15:27, or $\frac{5}{9}$, 5 to 9, 5:9

6. math teachers to Spanish teachers
$\frac{27}{18}$, 27 to 18, 27:18; or $\frac{3}{2}$, 3 to 2, 3:2

7. Spanish teachers to all teachers
$\frac{18}{60}$, 18 to 60, 18:60; or $\frac{3}{10}$, 3 to 10, 3:10

8. art and math teachers to Spanish teachers
$\frac{42}{18}$, 42 to 18, 42:18; or $\frac{7}{3}$, 7 to 3, 7:3

9. Thirty-two students are asked whether the school day should be longer. Twenty-four vote "no" and 8 vote "yes." Write the ratio of "no" votes to "yes" votes in simplest form.
3 to 1

10. A train car has 64 seats. There are 48 passengers on the train. Write the ratio of seats to passengers in simplest form.
4 to 3

11. Tell whose CD collection has the greater ratio of rock CDs to total CDs.
Glen

	Glen	Nina
Classical CDs	4	8
Rock CDs	9	12
Other CDs	5	7

/30

LESSON 5-1 Practice C: Ratios

A traveling theater company has 51 actors, 63 stagehands, and 27 set designers. Write each ratio in all three forms.

1. stagehands to set designers
$\frac{63}{27}$, 63 to 27, 63:27; or $\frac{7}{3}$, 7 to 3, 7:3

2. actors to stagehands
$\frac{51}{63}$, 51 to 63, 51:63; or $\frac{17}{21}$; 17 to 21, 17:21

3. set designers to actors
$\frac{27}{51}$, 27 to 51, 27:51; or $\frac{9}{17}$, 9 to 17, 9:17

4. set designers to stagehands and actors
$\frac{27}{114}$, 27 to 114, 27:114; or $\frac{9}{38}$, 9 to 38, 9:38

There are 18 monkeys, 6 gorillas, and 15 other apes in the Primate House at the zoo. Write different combinations of animals that have the following ratios.

5. 2:5 **gorillas to other apes**
6. 6:5 **monkeys to other apes**
7. 1:3 **gorillas to monkeys**
8. 6:7 **monkeys to gorillas and other apes**
9. 5:13 **other apes to all primates**
10. 8:5 **monkeys and gorillas to other apes**

11. The class library has 7 French books, 11 Spanish books, 19 art books, 3 Italian books, and 14 science books. What is the ratio of foreign language books to all books in simplest form? **7:18**

12. A market research company tests new commercials. Out of 96 people, 80 people prefer Commercial A and the rest prefer Commercial B. Write the ratio in simplest form of people who prefer Commercial A to Commercial B.
5 to 1 5:1 (ratio form)

13. Which grade has the greatest ratio of the number of students against wearing a uniform to the total number of students in that grade?
8th grade

School Uniform Survey	6th Grade	7th Grade	8th Grade
For	16	9	5
Against	20	24	21
No Opinion	4	3	1

/21

LESSON 5-1 Reteach: Ratios

A **ratio** is a comparison of two numbers.
Tamara has 2 dogs and 8 fish. The ratio of dogs to fish can be written in three different ways.

	Ratio	Ratio in simplest form
using the word *to*	2 to 8	1 to 4
using a colon (:)	2:8	1:4
writing a fraction	$\frac{2}{8}$	$\frac{1}{4}$

You can read the ratios as *2 to 8* or *1 to 4*.

In a basket of fruit, there are 8 apples, 3 bananas, and 5 oranges. Write each ratio in all three forms.

1. apples to bananas
There are **8** apples and **3** bananas. So, the ratio of apples to bananas is **8** to **3**, or **8 : 3**, or $\frac{8}{3}$

2. oranges to apples
There are **5** oranges and **8** apples. So, the ratio of oranges to apples is **5** to **8**, or **5 : 8**, or $\frac{5}{8}$

3. bananas to oranges
There are **3** bananas and **5** oranges. So, the ratio of bananas to oranges is **3** to **5**, or **3 : 5**, or $\frac{3}{5}$

4. apples to all pieces of fruit
There are **8** apples and **16** pieces of fruit in all. So, the ratio of apples to all pieces of fruit is **1** to **2**, or **1 : 2**, or $\frac{1}{2}$

A large bouquet of flowers is made up of 18 roses, 16 daisies, and 24 iris. Write each ratio in all three forms.

5. roses to iris
3 to 4; 3:4; $\frac{3}{4}$ 18:24

6. daisies to roses
8 to 9; 8:9; $\frac{8}{9}$ 16:18

7. iris to daisies
3 to 2; 3:2; $\frac{3}{2}$ 24:16

8. roses to all flowers
9 to 29; 9:29; $\frac{9}{29}$ 18:58

/40

LESSON 5-1 Reteach
Ratios (continued)

To compare ratios, write them as fractions with a common denominator. Then compare the numerators.

Tell whose bag has the greater ratio of solid marbles to striped marbles.

	Ken	Val
Solid	5	7
Striped	9	12

Step 1: Write the ratio of solid marbles to striped marbles for each bag. Write each ratio as a fraction.
Ken's Bag: $\frac{Solid}{Striped} = \frac{5}{9}$
Val's Bag: $\frac{Solid}{Striped} = \frac{7}{12}$

Step 2: Choose a common denominator. 36 is a common denominator for 9 and 12.

Step 3: Write each fraction using the common denominator.
$\frac{5}{9} = \frac{5 \times 4}{9 \times 4} = \frac{20}{36}$ $\frac{7}{12} = \frac{7 \times 3}{12 \times 3} = \frac{21}{36}$

Step 4: Compare the numerators. 20 < 21

Since $20 < 21, \frac{20}{36} < \frac{21}{36}$ and $\frac{5}{9} < \frac{7}{12}$. So, Val's bag has the greater ratio of solid marbles to striped marbles.

9. Tell whose bookshelf has the greater ratio of novels to biographies.

	Tina	Mark
Novels	3	5
Biographies	5	8

For Tina's bookshelf: $\frac{Novels}{Biographies} = \frac{3}{5}$
For Marks's bookshelf: $\frac{Novels}{Biographies} = \frac{5}{8}$

Write the fractions so they have a common denominator. Then compare the numerators.

$\frac{3}{5} = \frac{24}{40}, \frac{5}{8} = \frac{25}{40}, 24 < 25$

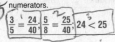's bookshelf has the greater ratio of novels to biographies.

LESSON 5-1 Challenge
The Golden Ratio

The Golden Ratio, which as also known as the Golden Section, was given its name because architects and artists throughout history have used it to produce shapes that are pleasing to look at. They have used this ratio as the ratio of length to width in various shapes.

The Golden Ratio can be represented by a line segment that meets these conditions:
- The line segment is divided into two sections of different lengths.
- The ratio of the longer section to the shorter section is equal to the ratio of the whole segment to the longer section

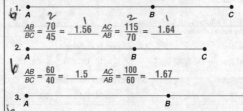

$\frac{longer? (AB)}{shorter? (BC)} = \frac{whole? (AC)}{longer? (AB)}$

Find each measurement to the nearest millimeter. Use your measurements to write the given ratio. Then write the ratio as a decimal rounded to the nearest hundredth.

1.
$\frac{AB}{BC} = \frac{70}{45} = 1.56$ $\frac{AC}{AB} = \frac{115}{70} = 1.64$

2.
$\frac{AB}{BC} = \frac{60}{40} = 1.5$ $\frac{AC}{AB} = \frac{100}{60} = 1.67$

3.
$\frac{AB}{BC} = \frac{80}{50} = 1.6$ $\frac{AC}{AB} = \frac{130}{80} = 1.63$

4. In which exercise is the line segment divided so that it is closest to representing the golden ratio? Explain.

Exercise 3; the ratio of the longer section to the shorter section is closest to equaling the ratio of the whole segment to the longer section in this exercise.

LESSON 5-1 Problem Solving
Ratios

Write the correct answer.

1. The Rockport Diner has 8 seats at the counter and 32 seats at tables. Of these seats, 16 are taken. Write the ratio of seats taken to empty seats in simplest form three ways.

   2 to 3; 2:3; $\frac{2}{3}$

2. During the 2001 WNBA season, the Los Angeles Sparks had 28 wins and only 4 losses. Write the ratio of wins to games played in simplest form three ways.

   7 to 8; 7:8; $\frac{7}{8}$

3. For every 300 people surveyed in 2002, 186 said their favorite Winter Olympic sport was figure skating. Write this ratio in simplest form three ways.

   50 to 31; 50:31; $\frac{50}{31}$

4. In 2004, George W. Bush received 286 electoral votes, and John Kerry received 251, and 1 elector voted for John Edwards. Write the ratio of Bush's electoral votes to total electoral votes in simplest form three ways.

   143 to 269; 143:269; $\frac{143}{269}$

Choose the letter for the best answer.

5. There are 62 girls in the seventh grade and 58 boys in the eighth grade. Each grade has 120 students. Which statement correctly compares the ratios of boys to girls in each grade?
   A The eighth-grade ratio is greater.
   B The seventh-grade ratio is greater.
   C The eighth-grade ratio is lesser.
   D Both ratios are equal.

6. Matt has 6 video racing games and 8 video sports games. Which ratio is the ratio of racing games to total video games in simplest form?
   F $\frac{3}{4}$ H $\frac{4}{3}$
   G $\frac{3}{7}$ J $\frac{4}{7}$

7. Which player has the greatest ratio of baskets to total shots?
   A Marisol
   B Nina
   C Joanne
   D Talia

	Baskets	Missed Shots
Marisol	8	8
Nina	7	5
Joanne	2	4
Talia	5	3

LESSON 5-1 Reading Strategies
Build Vocabulary

A ratio is a way to compare two numbers.

●●●○○

The ratio of black counters to white counters is three to two.

There are three ways to write ratios:

3 to 2 ← read "3 to 2"
3:2 ← read "3 to 2"
$\frac{3}{2}$ ← read "3 to 2"

Now compare the white counters to the black counters.

1. Use a fraction to write this ratio.
   $\frac{2}{3}$

2. Write the ratio two other ways.
   2 to 3; 2:3

A team has 5 boys and 4 girls. Use this information to complete Exercises 3 – 6.

3. Write a ratio as a fraction to compare the number of boys to the number of girls.
   $\frac{5}{4}$

4. Write this ratio two other ways.
   5 to 4; 5:4

5. Use division to compare the number of girls to the total number of players on the team.
   $\frac{4}{9}$

6. Write this ratio two other ways.
   4 to 9; 4:9

LESSON 5-1 Puzzles, Twisters & Teasers
Find Your Ratio

Use the letters in the word RATIO to write a fraction for each ratio in simplest form.

RATIO

Fraction Box
A = $\frac{1}{7}$
D = $\frac{3}{5}$
E = $\frac{2}{5}$
G = 1
H = $\frac{1}{11}$
I = $\frac{2}{11}$
L = $\frac{2}{7}$
M = $\frac{5}{7}$
O = $\frac{2}{5}$
R = $\frac{6}{11}$
S = $\frac{4}{7}$
T = $\frac{4}{11}$
Y = $\frac{7}{11}$

1. number of Ts:number of Rs $\frac{1}{1}$ or 1
2. consonants:vowels $\frac{2}{3}$
3. vowels:total number of letters $\frac{3}{5}$
4. consonants:total number of letters $\frac{2}{5}$

Next, use the letters in the word MATHEMATICS to write the fraction for each ratio in simplest form.

5. consonants:total number of letters $\frac{7}{11}$
6. vowels:consonants $\frac{4}{7}$
7. number of Ms:total number of letters $\frac{2}{11}$
8. number of Hs:total number of letters $\frac{1}{11}$
9. number of Ms and Ts:total number of letters $\frac{4}{11}$

Find the letters that match your answers in the Fraction Box. To solve the riddle, write the letters on the blanks that correspond to the problem numbers.

What do you need to spot an iceberg 20 miles away?

G O O D
1 2 2 3

E Y E S I G H T
4 5 4 6 7 1 8 9

LESSON 5-2 Practice A
Rates

1. To make 2 batches of brownies, Ed needs 4 eggs. How many eggs are needed per batch of brownies?
$\frac{4 \text{ eggs}}{2 \text{ batches}} = \frac{2 \text{ eggs}}{1 \text{ batch}}$
Ed needs __2__ eggs to make 1 batch of brownies.

2. Jenny drives 265 miles in 5 hours. What is her average rate of speed in miles per hour?
$\frac{265 \text{ miles}}{5 \text{ hours}} = \frac{53 \text{ miles}}{1 \text{ hour}}$
Jenny's average rate of speed is __53__ miles per hour.

3. A job pays $56 for 8 hours of work. How much money does the job pay per hour? __$7 per hour__

4. Ned scores 84 points in 6 games. How many points per game does Ned score? __14 points per game__

5. A 6-ounce blueberry muffin has 450 calories. How many calories are there per ounce? __75 calories per ounce__

6. A parking garage charges $21 for 6 hours. How much does the garage charge per hour? __$3.50 per hour__

7. The Rylands want to drive 360 miles in 8 hours. What should their average speed be in miles per hour? __45 miles per hour__

8. A plane travels 2,395 miles in 5 hours. What is the plane's average speed? __479 miles per hour__

9. A 16-ounce bottle of fruit punch costs $2.40. A 24-ounce bottle of fruit punch costs $3.84. Which size is the better buy?
$\frac{\$2.40}{16 \text{ oz}} = \frac{\$0.15}{1 \text{ oz}}$ $\frac{\$3.84}{24 \text{ oz}} = \frac{\$0.16}{1 \text{ oz}}$
The __16__-ounce bottle costs less per ounce.
So, the __16__-ounce bottle is the better buy.

LESSON 5-2 Practice B
Rates

1. A part-time job pays $237.50 for 25 hours of work. How much money does the job pay per hour?   **$9.50 per hour**

2. A class trip consists of 84 students and 6 teachers. How many students per teacher are there?   **14 students per teacher**

3. A factory builds 960 cars in 5 days. What is the average number of cars the factory produces per day?   **192 cars per day**

4. The Wireless Cafe charges $5.40 for 45 minutes of Internet access. How much money does The Wireless Cafe charge per minute?   **$0.12 per minute**

5. A bowler scores 3,152 points in 16 games. What is his average score in points per game?   **197 points per game**

6. Melissa drives 238 miles in 5 hours. What is her average rate of speed?   **47.6 miles per hour**

7. An ocean liner travels 1,233 miles in 36 hours. What is the ocean liner's average rate of speed?   **34.25 miles per hour**

8. A plane is scheduled to complete a 1,792-mile flight in 3.5 hours. In order to complete the trip on time, what should be the plane's average rate of speed?   **512 miles per hour**

9. The Nuthouse sells macadamia nuts in three sizes. The 12 oz jar sells for $8.65, the 16 oz jar sells for $10.99, and the 24 oz gift tin costs $16.99. Which size is the best buy?   **the 16 oz jar**

10. Nina paid $37.57 for 13 gallons of gas. Fred paid $55.67 for 19 gallons of gas. Eleanor paid $48.62 for 17 gallons of gas. Who got the best buy?   **Eleanor**

LESSON 5-2 Practice C
Rates

1. Maria earns $603.75 for 35 hours of work. What is her rate of pay per hour? __$17.25 per hour__

2. The Ranch House serves a 24 oz sirloin steak that has a total of 1,800 calories. How many calories per ounce does the steak have? __75 calories per ounce__

3. A volunteer stuffs 228 envelopes in an hour. What is the average number of envelopes the volunteer stuffs per minute? __3.8 envelopes per minute__

4. A freight train travels 1445 miles in 25 hours. What is the train's average rate of speed? __57.8 miles per hour__

5. June runs 600 yards in 2 min 5 sec. What is her average speed in yards per second? __4.8 yards per second__

6. A plane travels 294 miles in 45 minutes. What is its average speed in miles per hour? __392 miles per hour__

7. A pitcher's earned run average is the number of earned runs allowed per game, with a game defined as 9 innings. A pitcher allows 20 earned runs in 50 innings. What is the pitcher's earned run average? __3.6 earned runs per game__

Find each unit price. Then decide which is the better buy.

8. $\frac{\$6.48}{36 \text{ oz}}$ or $\frac{\$8.16}{48 \text{ oz}}$
$0.18 per oz; $0.17 per oz;
$\frac{\$8.16}{48 \text{ oz}}$ is the better buy.

9. $\frac{\$9.03}{7 \text{ ft}}$ or $\frac{\$15.84}{12 \text{ ft}}$
$1.29 per ft; $1.32 per ft;
$\frac{\$9.03}{7 \text{ ft}}$ is the better buy.

10. $\frac{\$25.35}{6.5 \text{ lb}}$ or $\frac{\$31.60}{8 \text{ lb}}$
$3.90 per lb; $3.95 per lb;
$\frac{\$25.35}{6.5 \text{ lb}}$ is the better buy.

11. $\frac{\$9.16}{0.4 \text{ m}}$ or $\frac{\$13.20}{0.6 \text{ m}}$
$22.90 per m; $22 per m;
$\frac{\$13.20}{0.6 \text{ m}}$ is the better buy.

LESSON 5-2 Reteach
Rates

A **rate** is a ratio that compares two different kinds of quantities or measurements. Rates can be simplified. Rates sometimes use the words *per* and *for* instead of *to*, such as 55 miles per hour and 3 tickets for $1.

The scale on a map might be 3 inches equals 60 miles. Simplify by dividing both numerator and denominator by the same number.

$$\frac{3 \text{ inches}}{60 \text{ miles}} = \frac{3 \div 3}{60 \div 3} = \frac{1 \text{ inch}}{20 \text{ miles}}, \text{ or 1 inch to 20 miles}$$

A **unit rate** is a rate per 1 unit. So, in a unit rate, the denominator is always 1.

Miguel can type 180 words in 4 minutes.

$$\underbrace{\frac{180 \text{ words}}{4 \text{ minutes}}}_{\text{rate}} = \frac{180 \div 4}{4 \div 4} = \underbrace{\frac{45 \text{ words}}{1 \text{ minute}}}_{\text{unit rate}}, \text{ or } \underbrace{45 \text{ words per minute}}_{\text{word form}}$$

Find each unit rate. Write in both fraction and word form.

1. Film costs $7.50 for 3 rolls.
$\frac{\$7.50}{3 \text{ rolls}} = \frac{7.50 \div 3}{3 \div 3} = \frac{\$2.50}{1 \text{ roll}}$
Word form: __$2.50 per roll__

2. Drive 288 miles on 16 gallons of gas.
$\frac{288 \text{ mi}}{16 \text{ gal}} = \frac{288 \div 16}{16 \div 16} = \frac{18 \text{ mi}}{1 \text{ gal}}$
Word form: __18 miles per gallon__

3. Earn $52 for 8 hours of work.
$\frac{\$52.00}{8 \text{ hr}} = \frac{52 \div 8}{8 \div 8} = \frac{\$6.50}{1 \text{ hr}}$
Word form: __$6.50 per hour__

4. Use 5 quarts of water for every 2 pounds of chicken.
$\frac{5 \text{ qt}}{2 \text{ lb}} = \frac{5 \div 2}{2 \div 2} = \frac{2.5 \text{ qt}}{1 \text{ lb}}$
Word form: __2.5 quarts per pound__

5. Snowfall of 12 inches in 4 hours
$\frac{3 \text{ in.}}{1 \text{ hr}}$; 3 inches per hour

6. 90 students and 5 teachers
$\frac{18 \text{ students}}{1 \text{ teacher}}$; 18 students per teacher

---

LESSON 5-2 Challenge
The Rate Maze

Contestants A–H must make their way through the maze to the Winners' Circle. To reach the Winners' Circle, each contestant must find a path from his or her current location through sections containing equivalent unit prices or rates. The contestants can move only through sections that share a corner.

Find each unit price or rate. Circle the two contestants who will *not* be able to get to the Winners' Circle.

LESSON 5-2 Problem Solving
Rates

Write the correct answer.

1. A truck driver drives from Cincinnati to Boston in 14 hours. The distance traveled is 840 miles. What is the truck driver's average rate of speed?
__60 miles per hour__

2. Melanie earns $97.50 in 6 hours. Earl earns $296.00 in 20 hours. Who earns a higher rate of pay per hour?
__Melanie__

3. Mr. Tanney buys a 10-trip train ticket for $82.50. Ms. Elmer buys an unlimited weekly pass for $100 and uses it for 12 trips during the week. Write the unit cost per trip for each person.
__Mr. Tanney, $8.25 per trip;__
__Ms. Elmer, $8.33 per trip__

4. Metropolitan Middle School has 564 students and 24 teachers. Eastern Middle School has 623 students and 28 teachers. Which school has the lower unit rate of students per teacher?
__Eastern Middle School__

Choose the letter for the best answer.

5. Which shows 20 pounds per 5 gallons as a unit rate?
A $\frac{20 \text{ lb}}{1 \text{ gal}}$
(B) $\frac{4 \text{ lb}}{1 \text{ gal}}$
C $\frac{5 \text{ lb}}{1 \text{ gal}}$
D $\frac{1 \text{ gal}}{4 \text{ lb}}$

6. What is the unit price of a 6-ounce tube of toothpaste that costs $3.75?
F $0.06
G $0.23
H $0.62
(J) $0.63

7. Max bought 16 gallons of gas for $40.64. Lydia bought 12 gallons of gas for $31.08. Kesia bought 18 gallons of gas for $45.72. Who got the best buy?
A Max got the best buy.
B Lydia got the best buy.
C Lydia and Kesia both paid the same rate, which is better than Max's rate.
(D) Max and Kesia both paid the same rate, which is better than Lydia's rate.

8. A pack of 12 8-ounce bottles of water costs $3.36. What is the unit cost per ounce of bottled water?
F $0.03 per ounce
(G) $0.04 per ounce
H $0.28 per bottle
J $0.42 per bottle

LESSON 5-2 Reading Strategies
Build Vocabulary

A **rate** is a special ratio that compares two values that are measured in different units.

- $8 for 2 pounds of beef
$\frac{\$8}{2 \text{ lb}}$
pounds compared to dollars

- 12 miles in 3 hours
$\frac{12 \text{ mi}}{3 \text{ h}}$
miles compared to hours

1. Is the ratio $\frac{5 \text{ hours}}{12 \text{ hours}}$ a rate? Explain.
__No; it does not compare values that are in different units.__

2. Is the ratio $\frac{50 \text{ yards}}{18 \text{ seconds}}$ a rate? Explain.
__Yes; it compares yards to seconds.__

3. What does the rate $\frac{375 \text{ miles}}{15 \text{ gallons}}$ compare?
__It compares miles to gallons.__

In a **unit rate**, the second quantity in the rate is 1 unit. To write a unit rate, write the rate as a fraction with a denominator of 1.

$$\underbrace{\frac{\$8}{2 \text{ lb}}}_{\text{rate}} = \frac{(\$8 \div 2)}{(2 \text{ lb} \div 2)} = \underbrace{\frac{\$4}{1 \text{ lb}}}_{\text{unit rate}}$$

Write *yes* or *no* to tell if each rate is a unit rate. If it is not a unit rate, write the unit rate.

4. $\frac{\$2.75}{1 \text{ h}}$ __Yes__

5. $\frac{100 \text{ mi}}{4 \text{ gal}}$ __No; $\frac{25 \text{ mi}}{1 \text{ gal}}$__

6. $\frac{35 \text{ lb}}{1 \text{ box}}$ __Yes__

7. $\frac{40 \text{ ft}}{5 \text{ s}}$ __No; $\frac{8 \text{ ft}}{1 \text{ s}}$__

LESSON 5-2 Puzzles, Twisters, & Teasers
At This Rate...

Draw a line to connect each rate to the equivalent unit rate.

1. $22.80 / 8 lb   18 ft / 1 s   O
2. 288 ft / 16 s $2.75 / 1 lb   I
3. $16.50 / 6 lb   17.6 ft / 1 s   M
4. 792 ft / 45 s $0.05 / 1 min   N
5. $3.74 / 68 min $2.85 / 1 lb   K
6. $4.05 / 75 min $0.06 / 1 min   A

Next, choose the best buy. Circle the letter next to your answer.

7. E. $3.33 / 4.5 lb   L. $2.28 / 3 lb   (C.) $2.92 / 4 lb
8. (O) $45.36 / 21 gal [O] B. $38.97 / 18 gal   T. $57.25 / 25 gal
9. S. $2.16 / 36 min   (A.) $3.30 / 60 min [A] G. $5.04 / 72 min
10. P. $34.32 / 8 yd   W. $49.92 / 12 yd   (R.) $65.44 / 16 yd

Write the letters that are next to your answers above the problem numbers to solve the riddle.

Where do tadpoles hang their coats?

 I   N   A
 3.   6.   5.

 C   R   O   A   K   R   O   O   M
 7. 10.   2.   9.   1. 10.   2.   8.   4.

LESSON 5-3 Practice A
Slope and Rates of Change

Choose the letter for the best answer.

1. What is the slope of the line?
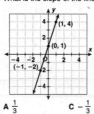

   A $\frac{1}{3}$ C $-\frac{1}{3}$
   (B) 3   D -3

2. What is the slope of the line?
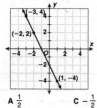

   A $\frac{1}{2}$ C $-\frac{1}{2}$
   B 2   (D) -2

Use the given slope and point to graph each line.

3. slope = 2; (1, −2)

4. slope = −1; (2, 0)

Tell whether each graph shows a constant or variable rate of change.

5.
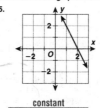
constant

6.
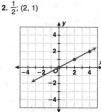
variable

LESSON 5-3 Practice B
Slope and Rates of Change

Tell whether the slope is positive or negative. Then find the slope.

1.
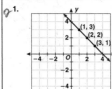
negative; −1

2.
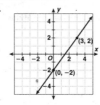
positive; $\frac{4}{3}$

Use the given slope and point to graph each line.

3. $-\frac{1}{2}$; (2, 4)
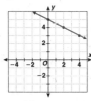

4. $\frac{1}{3}$; (−1, −2)
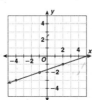

Tell whether each graph shows a constant or variable rate of change.

5.
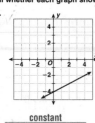
constant

6.
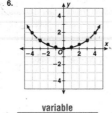
variable

LESSON 5-3 Practice C
Slope and Rates of Change

Use the given slope and point to graph each line.

1. $-\frac{1}{3}$; (−3, 5)

2. $\frac{1}{2}$; (2, 1)

Tell whether each graph shows a constant or variable rate of change.

3.

variable

4.

constant

5. The graph at the right shows the cost per pound of buying grapes.
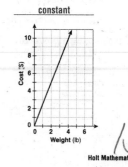

   a. Is the cost per pound a constant or a variable rate?
      constant

   b. What is the cost per pound of grapes?
      $2.50

LESSON 5-3 Reteach
Slope and Rates of Change

The **slope** of a line is a ratio that measures the steepness of that line. The sign of the slope tells whether the line is rising or falling.

You can find the slope of a line by comparing any two points on that line. Find the slope of the line between (2, 1) and (4, 4).

slope = $\frac{rise}{run}$ = $\frac{up (+) \text{ or down } (-)}{right (+) \text{ or left } (-)}$

= $\frac{4-1}{4-2} = \frac{3}{2}$

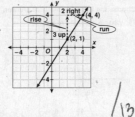

Use the graph to complete the statements.

1. You have to move up __3__ and right __2__ to go from one point to the other.

   slope = $\frac{rise}{run}$ = $\frac{up (+) \text{ or down } (-)}{right (+) \text{ or left } (-)}$ = $\frac{3}{2}$

2. The slope of the line is $\frac{3}{2}$.

Find the slope.

3.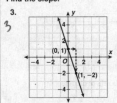

4.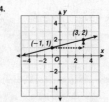

slope = $\frac{up \text{ or down}}{right \text{ or left}} = \frac{-3}{1} = -3$

slope = $\frac{up \text{ or down}}{right \text{ or left}} = \frac{1}{4}$

5. Explain how you can use the point (2, 3) and the slope of 2 to draw a line.

   Possible answer: Graph the point (2, 3). The slope is 2, or $\frac{2}{1}$. So, move 2 units up and 1 unit right, and mark a point. Then draw a line through the two points.

LESSON 5-3 Challenge
Interpret Linear Equations

A linear equation is an equation whose graph is a line. You can use a linear equation to find the slope of a line. You can also use a linear equation to find the y-intercept of a line. The **y-intercept** is the point where a line crosses the y-axis.

When the equation of the line is in the form $y = mx + b$, you can use m to identify the slope of the line and b to identify the y-intercept.

$y = mx + b$
$y = -4x + 2$
The slope is -4 and the y-intercept is 2. The coordinates of the y-intercept are (0, 2).

$y = mx + b$
$y = \left(\frac{2}{3}\right)x - 1$
The slope is $\frac{2}{3}$ and the y-intercept is -1. The coordinates of the y-intercept are (0, -1).

Find the slope and the coordinates of the y-intercept for each equation.

1. $y = 8x + 16$ 8; (0, 16)

2. $y = -3x + 4$ -3; (0, 4)

3. $y = \left(\frac{2}{3}\right)x - 4$ $\frac{2}{3}$; (0, -4)

4. $y = -\left(\frac{1}{2}\right)x + \frac{3}{4}$ $-\frac{1}{2}$; (0, $\frac{3}{4}$)

5. $y = 9x$ 9; (0, 0)

6. $y = -5x - 6$ -5; (0, -6)

7. $y = -x - 3\frac{1}{3}$ -1; (0, $-3\frac{1}{3}$)

8. $y = \frac{2}{3}x + 4$ $\frac{2}{3}$; (0, 4)

9. $y = \left(-\frac{1}{8}\right)x$ $-\frac{1}{8}$; (0, 0)

Rewrite each equation so that it is in the form $y = mx + b$. Then find the slope and the y-intercept.

10. $y - 1 = 2x + 2$ $y = 2x + 3$; 2; (0, 3)

11. $y + 4 = x + 1$ $y = x - 3$; 1; (0, -3)

12. $2y = 2x + 8$ $y = x + 4$; 1; (0, 4)

13. $3y = x - 6$ $y = \left(\frac{1}{3}\right)x - 2$; $\frac{1}{3}$; (0, -2)

LESSON 5-3 Problem Solving
Slope and Rates of Change

Write the correct answer.

1. How much does Jerry earn per hour?
   $8

2. What is the slope of the graph that represents Daniel's rate of pay? How much does Daniel earn per hour?
   6; $6

3. Jerry and Daniel each worked 10 hours this week. How much more than Daniel did Jerry earn?
   $20

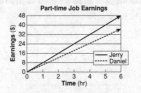

4. For more than 10 hours of work, the rate of pay is 1.5 times that shown in the graph. How much would Jerry earn by working 14 hours?
   $128

Choose the letter of the best answer.

The graph shows the changing height of two rockets over time. Use the graph to solve problems 5 and 6.

5. Which statement is true?
   A Both Rocket A and Rocket B have a constant rate of change in height.
   B Both Rocket A and Rocket B have a variable rate of change in height.
   C Rocket A has a variable rate of change in height, but Rocket B does not.
   D Rocket B has a variable rate of change in height, but Rocket A does not.

6. How fast is the height of rocket B increasing?
   F 5 feet per second
   G 10 feet per second
   H 20 feet per second
   J 40 feet per second

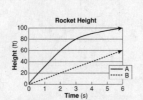

7. Jamaal plotted the point (1, -2). Then he used the slope $-\frac{2}{3}$ to find another point on the line. Which point could be the point that Jamaal found?
   A (-1, -1)
   B (4, 0)
   C (2, -3)
   D (4, -4)

LESSON 5-3 Reading Strategies
Compare and Contrast

The graph of Lupe's savings plan shows that she saves at a constant rate of $5 each week.

A graph with a **constant rate of change** is a line graph.

The graph of Alok's plan shows the amount he saves is different from one week to the next.

Compare the two graphs to answer the following questions.

1. How is Alok's savings plan different than Lupe's?
   Lupe saves the same amount each week, but Alok does not.

2. How do the graphs show a difference in their savings plans?
   Lupe's graph is a straight line; Alok's graph is not.

3. Compare the totals that Alok and Lupe had saved after 4 weeks.
   They had both saved $20.

4. Why is one graph a straight line and the other is not?
   The constant rate gives a graph of a straight line; the changing rate does not.

LESSON 5-3 Puzzles, Twisters, & Teasers
Slippin' and Slopin'!

Across
4. The _____ of a line is a measure of its steepness and is the ratio of rise to run.
5. A line that slopes upward has a _____ slope.
6. When you know the slope and any one point on a line, you can graph the _____
7. A graph that is a line has a _____ rate of change.

Down
1. A graph that is a curve has a _____ rate of change.
2. The _____ describes the difference between the y-coordinates of two points.
3. A line that slopes downward has a _____ slope.
5. Given one _____ and the slope, you can graph a line.

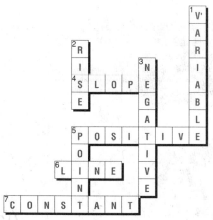

Crossword answers: 1V-VARIABLE, 2R-RISE, 3N-NEGATIVE, 4S-SLOPE, 5P-POSITIVE, 6L-LINE, 7C-CONSTANT

LESSON 5-4 Practice A
Identifying and Writing Proportions

Write the ratios in simplest form. Determine if the ratios are proportional by comparing them.

1. $\frac{1}{4}, \frac{3}{12}$ 2. $\frac{2}{3}, \frac{6}{9}$ 3. $\frac{4}{5}, \frac{15}{20}$

$\frac{1}{4}, \frac{1}{4}$; yes $\frac{2}{3}, \frac{2}{3}$; yes $\frac{4}{5}, \frac{3}{4}$; no

4. $\frac{3}{6}, \frac{6}{12}$ 5. $\frac{5}{6}, \frac{16}{18}$ 6. $\frac{2}{5}, \frac{6}{15}$

$\frac{1}{2}, \frac{1}{2}$; yes $\frac{5}{6}, \frac{8}{9}$; no $\frac{2}{5}, \frac{2}{5}$; yes

7. $\frac{1}{3}, \frac{3}{9}$ 8. $\frac{4}{6}, \frac{7}{12}$ 9. $\frac{3}{4}, \frac{18}{24}$

$\frac{1}{3}, \frac{1}{3}$; yes $\frac{2}{3}, \frac{7}{12}$; no $\frac{3}{4}, \frac{3}{4}$; yes

10. $\frac{2}{3}, \frac{9}{15}$ 11. $\frac{2}{4}, \frac{9}{20}$ 12. $\frac{3}{5}, \frac{15}{25}$

$\frac{2}{3}, \frac{3}{5}$; no $\frac{1}{2}, \frac{9}{20}$; no $\frac{3}{5}, \frac{3}{5}$; yes

Find an equivalent ratio. Then write the proportion. Answers may vary. Possible answers given.

13. $\frac{1}{2}$ 14. $\frac{3}{4}$ 15. $\frac{5}{8}$

$\frac{1}{2} = \frac{5}{10}$ $\frac{3}{4} = \frac{6}{8}$ $\frac{5}{8} = \frac{25}{40}$

16. $\frac{4}{6}$ 17. $\frac{1}{7}$ 18. $\frac{10}{25}$

$\frac{4}{6} = \frac{2}{3}$ $\frac{1}{7} = \frac{5}{35}$ $\frac{10}{25} = \frac{2}{5}$

LESSON 5-4 Practice B
Identifying and Writing Proportions

Determine whether the ratios are proportional.

1. $\frac{3}{4}, \frac{24}{32}$ yes   2. $\frac{5}{6}, \frac{15}{18}$ yes   3. $\frac{10}{12}, \frac{20}{32}$ no

4. $\frac{7}{10}, \frac{22}{30}$ no   5. $\frac{9}{6}, \frac{21}{14}$ yes   6. $\frac{7}{9}, \frac{24}{27}$ no

7. $\frac{4}{10}, \frac{6}{15}$ yes   8. $\frac{7}{12}, \frac{13}{20}$ no   9. $\frac{4}{9}, \frac{6}{12}$ no

10. $\frac{7}{8}, \frac{14}{16}$ yes   11. $\frac{9}{10}, \frac{45}{50}$ yes   12. $\frac{3}{7}, \frac{10}{21}$ no

Find a ratio equivalent to each ratio. Then use the ratios to write a proportion. Answers may vary. Possible answers given.

13. $\frac{7}{9}$ 14. $\frac{11}{12}$ 15. $\frac{14}{15}$

$\frac{7}{9} = \frac{70}{90}$ $\frac{11}{12} = \frac{110}{120}$ $\frac{14}{15} = \frac{28}{30}$

16. $\frac{35}{55}$ 17. $\frac{14}{10}$ 18. $\frac{25}{18}$

$\frac{35}{55} = \frac{70}{110}$ $\frac{14}{10} = \frac{7}{5}$ $\frac{25}{18} = \frac{100}{72}$

LESSON 5-4 Practice C
Identifying and Writing Proportions

Determine whether the ratios are proportional.

1. $\frac{7}{11}, \frac{42}{60}$ no   2. $\frac{10}{18}, \frac{38}{72}$ no   3. $\frac{18}{28}, \frac{27}{42}$ yes

4. $\frac{6}{14}, \frac{15}{35}$ yes   5. $\frac{9}{24}, \frac{16}{40}$ no   6. $\frac{12}{39}, \frac{20}{65}$ yes

Find a ratio equivalent to each ratio. Then use the ratios to write a proportion. Answers may vary. Possible answers given.

7. $\frac{7}{31}$ 8. $\frac{24}{51}$ 9. $\frac{6}{29}$

$\frac{7}{31} = \frac{21}{93}$ $\frac{24}{51} = \frac{48}{102}$ $\frac{6}{29} = \frac{24}{116}$

10. $\frac{14}{23}$ 11. $\frac{17}{39}$ 12. $\frac{25}{32}$

$\frac{14}{23} = \frac{42}{69}$ $\frac{17}{39} = \frac{34}{78}$ $\frac{25}{32} = \frac{100}{128}$

Complete each table of equivalent ratios.

13. 4 CDs to 10 tapes

CDs	2	4	10	12	28
Tapes	5	10	25	30	70

14. 9 triangles per 6 circles

Triangles	3	9	12	30	75
Circles	2	6	8	20	50

Find two ratios equivalent to each given ratio. Answers may vary. Possible answers given.

15. 10:21   20:42; 40:84   16. 15:8   30:16; 150:80

17. $\frac{5}{9}$ $\frac{10}{18}, \frac{20}{36}$ 18. $\frac{24}{11}$ $\frac{48}{22}, \frac{96}{44}$

Holt Mathematics   83

LESSON 5-4 Reteach
Identifying and Writing Proportions

Two ratios that are equal form a **proportion**. To determine whether two ratios are proportional, find the cross products of the ratios. If the cross products are equal, then the ratios are proportional.

If $a \cdot d = b \cdot c$, then $\frac{a}{b} = \frac{c}{d}$.

Are $\frac{6}{9}$ and $\frac{8}{12}$ proportional?
Find the cross products.
$6 \cdot 12 = 72$ and $9 \cdot 8 = 72$
Since the cross products are equal, $\frac{6}{9}$ and $\frac{8}{12}$ are proportional.
So, $\frac{6}{9} = \frac{8}{12}$.

Are $\frac{4}{10}$ and $\frac{3}{8}$ proportional?
Find the cross products.
$4 \cdot 8 = 32$ and $10 \cdot 3 = 30$
Since the cross products are not equal, $\frac{4}{10}$ and $\frac{3}{8}$ are not proportional.
So, $\frac{4}{10} \neq \frac{3}{8}$.

Find the cross products to determine if the ratios are proportional.

1. $\frac{15}{21}, \frac{5}{7}$ $15 \cdot \underline{7} = \underline{105}$ $21 \cdot \underline{5} = \underline{105}$
   Are the ratios proportional? **yes**

2. $\frac{6}{9}, \frac{9}{15}$ $6 \cdot \underline{15} = \underline{90}$ $9 \cdot 9 = \underline{81}$
   Are the ratios proportional? **no**

3. $\frac{15}{6}, \frac{9}{4}$ **60, 54, no**

4. $\frac{12}{24}, \frac{5}{10}$ **120, 120, yes**

5. $\frac{20}{12}, \frac{15}{9}$ **180, 180, yes**

You can write a proportion from a given ratio. Multiply or divide the numerator and denominator of the ratio by the same number.

$\frac{9}{12} = \frac{9 \div 3}{12 \div 3} = \frac{3}{4}$ So, $\frac{9}{12} = \frac{3}{4}$. $\frac{9}{12} = \frac{9 \cdot 4}{12 \cdot 4} = \frac{36}{48}$ So, $\frac{9}{12} = \frac{36}{48}$.

Find an equivalent ratio. Then write the proportion. Possible answers given.

6. $\frac{6}{10}$ $\frac{6}{10} = \frac{3}{5}$

7. $\frac{10}{15}$ $\frac{10}{15} = \frac{20}{30}$

8. $\frac{18}{24}$ $\frac{18}{24} = \frac{3}{4}$

LESSON 5-4 Challenge
What's in the Set?

The ratios below describe a set of polygons. Use the ratios to find the number of each type of polygon in the set. All ratios are written in simplest form.

- Equilateral triangles to isosceles triangles is 2 to 3
- Isosceles triangles to scalene triangles is 6:5
- Triangles to rectangles is $\frac{5}{4}$
- Parallelograms to triangles is $\frac{4}{3}$

1. What is the fewest number of each polygon that can be in the set?
   a. Equilateral triangles __4__
   b. Isosceles triangles __6__
   c. Scalene triangles __5__
   d. All triangles __15__
   e. Rectangles __12__
   f. Parallelograms __20__

2. Explain the order in which you determined the number of each polygon in the set.
   Answers will vary. Possible answer: Find the number of triangles first. There must be at least 5 scalene triangles. Then use the ratios.

3. Other sets of polygons with the same ratios are possible. Find the number of polygons in another set that has the same ratios.
   a. Equilateral triangles __12__   Answers will vary. Possible answers given.
   b. Isosceles triangles __18__
   c. Scalene triangles __15__
   d. All triangles __45__
   e. Rectangles __36__
   f. Parallelograms __60__

LESSON 5-4 Problem Solving
Identifying and Writing Proportions

Write the correct answer.

1. Jeremy earns $234 for 36 hours of work. Miguel earns $288 for 40 hours of work. Are the pay rates of these two people proportional? Explain.
   No; $\frac{234}{36} = \frac{6.5}{1}$, $\frac{288}{40} = \frac{7.2}{1}$, $\frac{6.5}{1} \neq \frac{7.2}{1}$

2. Marnie bought two picture frames. One is 5 inches by 8 inches. The other is 15 inches by 24 inches. Are the ratios of length to width proportional for these frames? Explain.
   Yes; $\frac{5}{8} = \frac{15}{24}$

3. The ratio of adults to children at a picnic is 4 to 5. The total number of people at the picnic is between 20 and 30. Write an equivalent ratio to find how many adults and children are at the picnic.
   $\frac{12}{15}$; 12 adults and 15 children

4. A recipe for fruit punch calls for 2 cups of pineapple juice for every 3 cups of orange juice. Write an equivalent ratio to find how many cups of pineapple juice should be used with 12 cups of orange juice.
   $\frac{8}{12}$; 8 cups of pineapple juice

Choose the letter for the best answer.

5. A clothing store stocks 5 blouses for every 3 pairs of pants. Which ratio is proportional for the number of pairs of pants to blouses?
   A 15:9   **C 12:20**
   B 3:8   D 18:25

6. To make lemonade, you can mix 4 teaspoons of lemonade powder with 16 ounces of water. What is the ratio of powder to water?
   F 4:32   H 24:64
   G 32:8   **J 32:128**

7. The town library is open 4 days per week. Suppose you use the ratio of days open to days in a week to find the number of days open in 5 weeks. What proportion could you write?
   A $\frac{4}{7} = \frac{20}{25}$ **C $\frac{4}{7} = \frac{20}{28}$**
   B $\frac{7}{4} = \frac{21}{12}$ D $\frac{4}{7} = \frac{20}{35}$

8. At a factory, the ratio of defective parts to total number of parts is 3:200. Which is an equivalent ratio?
   F 6:1000   H 30:1000
   G 150:10,000   J 1,000:10,000

LESSON 5-4 Reading Strategies
Compare and Contrast

A **proportion** is two equal ratios.
Here are two ratios: $\frac{6}{8}$ and $\frac{9}{12}$.

To find out if they are equal, reduce ratios to simplest form.

$\frac{6}{8} = \frac{3}{4}$ $\frac{9}{12} = \frac{3}{4}$

$\frac{6}{8}$ and $\frac{9}{12}$ are **equal ratios**. They form a proportion.
Read: "6 is to 8 as 9 is to 12."
Compare these two ratios: $\frac{4}{7}$ and $\frac{5}{9}$.
These ratios are in simplest form, but they are *not equal*.

$\frac{4}{7} \neq \frac{5}{9}$

$\frac{4}{7}$ and $\frac{5}{9}$ are not equal ratios. They do *not* form a proportion.

Use the ratios $\frac{4}{6}$ and $\frac{1}{3}$ to answer Exercises 1–3.

1. Reduce $\frac{4}{6}$ to simplest form. $\frac{2}{3}$

2. Compare $\frac{4}{6}$ and $\frac{1}{3}$. Are they equal ratios?
   no

3. Do these two ratios form a proportion? Why or why not?
   No, because the ratios are not equal.

Use the ratios $\frac{2}{5}$ and $\frac{4}{10}$ for Exercises 4–6.

4. Reduce $\frac{4}{10}$ to simplest form. $\frac{2}{5}$

5. Compare $\frac{2}{5}$ and $\frac{4}{10}$. Are they equal ratios?
   yes

6. Do $\frac{2}{5}$ and $\frac{4}{10}$ form a proportion? Why or why not?
   Yes, because the ratios are equal.

LESSON 5-4 Puzzles, Twisters & Teasers
Draw the Line!

Draw lines to connect the proportional ratios. Then start with number 1 and find the letter next to the answer. Use the letters with each correct answer to solve the riddle.

1. $\frac{8}{14}$
2. $\frac{2}{4}$
3. $\frac{18}{24}$
4. $\frac{10}{6}$
5. $\frac{6}{21}$
6. $\frac{6}{20}$
7. $\frac{1}{3}$
8. $\frac{1}{5}$
9. $\frac{5}{3}$
10. $\frac{10}{12}$

$\frac{20}{12}$ T
$\frac{3}{9}$ E
$\frac{5}{6}$ R
$\frac{2}{7}$ S
$\frac{2}{10}$ D
$\frac{3}{10}$ Y
$\frac{25}{15}$ A
$\frac{3}{4}$ I
$\frac{12}{24}$ H
$\frac{4}{7}$ U

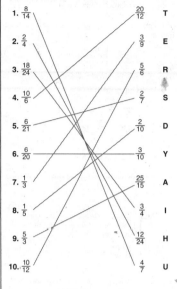

Why do soccer players do well in math?

T H E Y   U S E
4 2 7 6   1 5 7

T H E I R   H E A D S
4 2 7 3 10   2 7 9 8 5

LESSON 5-5 Practice A
Solving Proportions

Find the cross products.

1. $\frac{1}{2} = \frac{x}{8}$ 8 and 2x
2. $\frac{a}{6} = \frac{7}{9}$ 9a and 42
3. $\frac{5}{b} = \frac{8}{10}$ 50 and 8b

Use cross products to solve each proportion.

4. $\frac{2}{5} = \frac{x}{10}$ $x = 4$
5. $\frac{1}{3} = \frac{z}{15}$ $z = 5$
6. $\frac{3}{8} = \frac{s}{16}$ $s = 6$
7. $\frac{4}{r} = \frac{1}{4}$ $r = 16$
8. $\frac{10}{h} = \frac{5}{6}$ $h = 12$
9. $\frac{1}{d} = \frac{4}{12}$ $d = 3$
10. $\frac{w}{9} = \frac{6}{18}$ $w = 3$
11. $\frac{t}{8} = \frac{3}{4}$ $t = 6$
12. $\frac{k}{5} = \frac{9}{15}$ $k = 3$
13. $\frac{3}{6} = \frac{1}{f}$ $f = 2$
14. $\frac{2}{7} = \frac{6}{d}$ $d = 21$
15. $\frac{2}{9} = \frac{4}{c}$ $c = 18$
16. $\frac{a}{20} = \frac{15}{10}$ $a = 30$
17. $\frac{21}{k} = \frac{7}{4}$ $k = 12$
18. $\frac{3}{8} = \frac{n}{40}$ $n = 15$

19. Yolanda drove 50 miles in 2 hours at a constant speed. Use a proportion to find how long it would take her to drive 150 miles at the same speed.

$\frac{50}{2} = \frac{150}{x}$; $x = 6$ hours

LESSON 5-5 Practice B
Solving Proportions

Use cross products to solve each proportion.

1. $\frac{2}{5} = \frac{x}{35}$ $x = 14$
2. $\frac{7}{r} = \frac{1}{4}$ $r = 28$
3. $\frac{k}{75} = \frac{9}{15}$ $k = 45$
4. $\frac{1}{3} = \frac{z}{27}$ $z = 9$
5. $\frac{2}{11} = \frac{12}{d}$ $d = 66$
6. $\frac{24}{s} = \frac{4}{12}$ $s = 72$
7. $\frac{w}{42} = \frac{6}{7}$ $w = 36$
8. $\frac{t}{54} = \frac{2}{9}$ $t = 12$
9. $\frac{3}{8} = \frac{a}{64}$ $a = 24$
10. $\frac{17}{34} = \frac{7}{f}$ $f = 14$
11. $\frac{15}{h} = \frac{5}{6}$ $h = 18$
12. $\frac{4}{15} = \frac{36}{c}$ $c = 135$
13. $\frac{z}{25} = \frac{12}{5}$ $z = 60$
14. $\frac{36}{k} = \frac{9}{4}$ $k = 16$
15. $\frac{5}{14} = \frac{n}{42}$ $n = 15$
16. $\frac{8}{9} = \frac{40}{m}$ $m = 45$
17. $\frac{7}{c} = \frac{63}{54}$ $c = 6$
18. $\frac{24}{21} = \frac{s}{35}$ $s = 40$
19. $\frac{e}{22} = \frac{6}{15}$ $e = 8.8$
20. $\frac{3}{v} = \frac{12}{17}$ $v = 4.25$
21. $\frac{5}{14} = \frac{4}{a}$ $a = 11.2$

22. Eight oranges cost $1.00. How much will 5 dozen oranges cost?
$7.50

23. A recipe calls for 2 eggs to make 10 pancakes. How many eggs will you need to make 35 pancakes?
7 eggs

LESSON 5-5 Practice C
Solving Proportions

Use cross products to solve each proportion.

1. $\frac{3}{7} = \frac{x}{49}$ $x = 21$
2. $\frac{4}{11} = \frac{z}{55}$ $z = 20$
3. $\frac{16}{9} = \frac{64}{a}$ $a = 36$
4. $\frac{13}{r} = \frac{1}{5}$ $r = 65$
5. $\frac{17}{41} = \frac{34}{f}$ $f = 82$
6. $\frac{k}{18} = \frac{11}{3}$ $k = 66$
7. $\frac{w}{39} = \frac{7}{13}$ $w = 21$
8. $\frac{7}{19} = \frac{t}{95}$ $t = 35$
9. $\frac{65}{j} = \frac{13}{17}$ $j = 85$
10. $\frac{15}{h} = \frac{5}{17}$ $h = 51$
11. $\frac{1.7}{3} = \frac{d}{21}$ $d = 11.9$
12. $\frac{5}{19} = \frac{35}{c}$ $c = 133$
13. $\frac{28}{9} = \frac{19.6}{m}$ $m = 6.3$
14. $\frac{e}{136} = \frac{13}{17}$ $e = 104$
15. $\frac{3.7}{3} = \frac{s}{21}$ $s = 25.9$

Arrange the four numbers to form a proportion that is true. Answers may vary. Possible answers given.

16. 40, 50, 80, 64 $\frac{40}{50} = \frac{64}{80}$
17. 7, 9, 56, 72 $\frac{7}{56} = \frac{9}{72}$
18. 50, 45, 15, 150 $\frac{15}{45} = \frac{50}{150}$

19. A farmer can harvest 54 pounds of corn from each acre in his field. How much corn can he get from 12.5 acres?
675 pounds

20. A recipe for lasagna calls for 3 pounds of tomatoes to serve 5 people. A caterer wants to make enough lasagna to serve 110 people. How many pounds of tomatoes does he need?
66 pounds

LESSON 5-5 Reteach
Solving Proportions

Solving a proportion is like solving an equation involving fractions.
- Multiply both sides of the equation by the denominator of the fraction containing the variable.
- If the variable is in the denominator, invert both fractions in the proportion.

$\frac{n}{7} = \frac{20}{28}$

$7 \cdot \frac{n}{7} = 7 \cdot \frac{20}{28}$

$n = \frac{7 \cdot 20}{28} = \frac{140}{28}$

$n = 5$

$\frac{12}{x} = \frac{9}{6}$ *Invert both fractions.*

$\frac{x}{12} = \frac{6}{9}$

$12 \cdot \frac{x}{12} = 12 \cdot \frac{6}{9}$

$x = \frac{12 \cdot 6}{9} = \frac{72}{9}$

$x = 8$

Solve the proportion.

1. $\frac{a}{2} = \frac{27}{18}$

   $2 \cdot \frac{a}{2} = \frac{27}{18} \cdot 2$

   $a = \frac{27 \cdot 2}{18}$

   $a = \frac{54}{18}$

   $a = 3$

2. $\frac{8}{12} = \frac{n}{9}$

   $9 \cdot \frac{8}{12} = \frac{n}{9} \cdot 9$

   $\frac{9 \cdot 8}{12} = n$

   $\frac{72}{12} = n$

   $n = 6$

3. $\frac{10}{t} = \frac{4}{6}$

   $\frac{t}{10} = \frac{6}{4}$

   $10 \cdot \frac{t}{10} = \frac{6}{4} \cdot 10$

   $t = \frac{6 \cdot 10}{4}$

   $t = \frac{60}{4}$

   $t = 15$

4. $\frac{x}{15} = \frac{8}{10}$

   $x = 12$

5. $\frac{7}{3} = \frac{w}{18}$

   $w = 42$

6. $\frac{3}{2} = \frac{15}{c}$

   $c = 10$

LESSON 5-5 Reteach
Solving Proportions (continued)

You can use proportions to solve word problems.

A fruit punch is made with 32 ounces of ginger ale for every 12 ounces of frozen orange juice concentrate. How much ginger ale should you use for 30 ounces of orange juice concentrate?

- Set up a proportion comparing the amounts of ginger ale to orange juice concentrate.
- The first ratio shows the given recipe for the fruit punch.
- The second ratio shows the unknown amount of ginger ale as the variable g.
- Then solve the proportion.

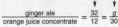

$\frac{\text{ginger ale}}{\text{orange juice concentrate}} = \frac{32}{12} = \frac{g}{30}$

$\frac{g}{30} = \frac{32}{12}$

$30 \cdot \frac{g}{30} = 30 \cdot \frac{32}{12}$

$g = \frac{30 \cdot 32}{12} = \frac{960}{12}$

$g = 80$

You should use 80 ounces of ginger ale for 30 ounces of frozen orange juice concentrate.

Solve.

7. Pecans cost $8.25 for 3 pounds. What is the cost of 5 pounds of pecans?

   $\frac{\text{dollars}}{\text{pounds}} = \frac{8.25}{3} = \frac{c}{5}$

   $c = $ **$13.75**

8. Mandy drove 90 miles in 2 hours at a constant speed. How long would it take her to drive 225 miles at the same speed?

   $\frac{\text{miles}}{\text{hours}} = \frac{90}{2} = \frac{225}{h}$

   $h = $ **5 hours**

9. Last week Geraldo bought 7 pounds of apples for $5.95. This week apples are the same price, and he buys 4 pounds. How much will he pay?

   $3.40

10. Aretha can type 55 words per minute. At that rate, how long will it take her to type a letter containing 935 words?

    17 minutes

LESSON 5-5 Challenge
Directly Speaking

Two quantities vary directly if they change in the same direction.
- If one quantity increases, then the other increases.
- If one quantity decreases, then the other decreases.

Here are some examples of direct variation.
- Amount bought and total cost
- Number of inches on a map and number of actual miles

If Farmer Jones plants 8 acres that produce 144 crates of melons, how many acres will produce 1,152 crates of melons?

Use the following proportion to solve.

$\frac{\text{acres for 144 crates}}{\text{144 crates}} = \frac{\text{acres for 1,152 crates}}{\text{1,152 crates}}$

$\frac{8}{144} = \frac{x}{1,152}$ $144x = 9,216$ $x = 64$

It will take 64 acres to produce 1,152 crates of melons.

Complete the proportion to find the answer.

		Do quantities increase or decrease?	Proportion	Answer
1.	If a furnace uses 40 gallons of oil in 8 days, how many gallons does it use in 10 days?	both increase	$\frac{40 \text{ gal}}{8 \text{ d}} = \frac{x \text{ gal}}{10 \text{ d}}$	50 gallons
2.	If 20 yards of wire weigh 80 pounds, what is the weight of 2 yards of the same wire?	both decrease	$\frac{20 \text{ yd}}{80 \text{ lb}} = \frac{2 \text{ yd}}{x \text{ lb}}$	8 pounds
3.	A cookie recipe calls for 2.5 cups of sugar and 4 eggs. If 6 eggs are used, how much sugar is needed?	both increase	$\frac{2.5 \text{ c}}{4} = \frac{x \text{ c}}{6}$	3.75 cups
4.	The scale on a map is 1 inch = 60 miles. What distance on the map represents 300 miles?	both increase	$\frac{1 \text{ in.}}{60 \text{ mi}} = \frac{x \text{ in.}}{300 \text{ mi}}$	5 inches
5.	At the market, 4 limes cost $1.50. How much will 10 limes cost?	both increase	$\frac{4}{\$1.50} = \frac{10}{x}$	$3.75

LESSON 5-5 Problem Solving
Solving Proportions

Write the correct answer.

1. Euros are currency used in several European countries. On one day in October 2005, you could exchange $3 for about 2.5 euros. How many dollars would you have needed to get 8 Euros?

   9.6, or $9.60

2. A 3-ounce serving of tuna fish provides 24 grams of protein. How many grams of protein are in a 10-ounce serving of tuna fish?

   80 grams

3. Hooke's law states that the distance a spring is stretched is directly proportional to the force applied. If 20 pounds of force stretches a spring 4 inches, how much will the spring stretch if 80 pounds of force is applied?

   16 inches

4. Beeswax used in making candles is produced by honeybees. The honeybees produce 7 pounds of honey for each pound of wax they produce. How many pounds of honey is produced if 145 pounds of beeswax?

   1,015 pounds of honey

Choose the letter for the best answer.

5. For every 5 books her students read, Mrs. Fenway gives them a free homework pass for 4 days. Juan has accumulated homework passes for 12 days so far. What proportion would you write to find how many books Juan has read?

   A $\frac{4}{12} = \frac{x}{5}$
   B $\frac{4}{5} = \frac{x}{12}$
   C $\frac{4}{5} = \frac{12}{x}$
   D $\frac{5}{12} = \frac{4}{x}$

6. In his last 13 times at bat in the township baseball league, Santiago got 8 hits. If he is at bat 65 times for the season, how many hits will he get if his average stays the same?

   F $\frac{8}{65} = \frac{x}{13}$
   G $\frac{x}{13} = \frac{13}{8}$
   H $\frac{8}{x} = \frac{65}{13}$
   J $\frac{8}{13} = \frac{x}{65}$

7. A 12-pack of 8-ounce juice boxes costs $5.40. How much would an 18-pack of juice boxes cost if it is proportionate in price?

   A $9.40
   B $8.10
   C $3.60
   D $12.15

8. Jeanette can swim 105 meters in 70 seconds. How far can she probably swim in 30 seconds?

   F 20 meters
   G 245 meters
   H 45 meters
   J 55 meters

LESSON 5-5 Reading Strategies
Draw a Conclusion

There is a quick method to check to see if two ratios are equal. It is called the **cross products** method. Follow these steps.

Step 1: Multiply factors that cross diagonally in the two ratios.

$\frac{4}{5} \times \frac{12}{15}$ ← 5×12
← 4×15

Step 2: If the products of cross factors are the same, the two ratios are equal.

The cross products of 60 are the same.

So, $\frac{4}{5}$ and $\frac{12}{15}$ are equal ratios. → $\frac{4}{5} \times \frac{12}{15}$ $\begin{matrix}60\\60\end{matrix}$

Two equal ratios form a **proportion**. $\frac{4}{5} = \frac{12}{15}$ is a proportion.

If cross products are not the same, the two ratios are not equal.

Is $\frac{4}{6} \stackrel{?}{=} \frac{5}{9}$?

$\frac{4}{6} \times \frac{5}{9}$ $5 \times 6 = 30$
$4 \times 9 = 36$

The cross products are not equal.

$\frac{4}{6} = \frac{5}{9}$ **is not** a proportion, so $\frac{4}{6} \ne \frac{5}{9}$.

Use the cross products method to tell whether the two ratios are equal. Show your work. Write yes or no.

1. $\frac{2}{4}$ and $\frac{6}{12}$ __yes__
2. $\frac{2}{5}$ and $\frac{5}{10}$ __no__
3. $\frac{6}{9}$ and $\frac{8}{12}$ __yes__
4. $\frac{3}{9}$ and $\frac{4}{10}$ __no__

5. How can you tell if two ratios form a proportion?

__If the cross products of two ratios are equal, the ratios form a proportion.__

LESSON 5-5 Puzzles, Twisters & Teasers
The Proper Proportions!

Decide whether each pair of ratios is a proportion. Circle the letter above your answer. Then start with number 1, and use the letters to solve the riddle.

1. $\frac{2}{5} = \frac{6}{15}$ (A) correct J incorrect
2. $\frac{6}{10} = \frac{36}{60}$ (L) correct P incorrect
3. $\frac{4}{7} = \frac{5}{6}$ K correct (L) incorrect
4. $\frac{4}{8} = \frac{12}{24}$ (T) correct M incorrect
5. $\frac{1}{3} = \frac{86}{255}$ N correct (H) incorrect
6. $\frac{39}{4} = \frac{121}{12}$ B correct (E) incorrect
7. $\frac{2}{15} = \frac{12}{90}$ (F) correct V incorrect
8. $\frac{18}{90} = \frac{1}{5}$ (A) correct C incorrect
9. $\frac{45}{9} = \frac{15}{3}$ (N) correct Z incorrect
10. $\frac{34}{6} = \frac{96}{16}$ L correct (S) incorrect
11. $\frac{3}{24} = \frac{4}{52}$ K correct (L) incorrect
12. $\frac{14}{20} = \frac{5}{8}$ G correct (E) incorrect
13. $\frac{2}{5} = \frac{3}{12}$ U correct (F) incorrect
14. $\frac{35}{4} = \frac{175}{20}$ (T) correct E incorrect

Why did it get hot after the baseball game?

__A__ __L__ __L__

__T__ __H__ __E__

__F__ __A__ __N__ __S__

__L__ __E__ __F__ __T__

LESSON 5-6 Practice A
Customary Measurements

Choose the letter of the best unit for each measurement.

1. The height of a house
   A inches (C) feet
   B pounds D miles

2. The weight of a letter
   (F) ounces H inches
   G pounds J tons

3. The capacity of a bathtub
   (A) gallons C fluid ounces
   B cups D pounds

4. The weight of a bowling ball
   F ounces H tons
   G fluid ounces (J) pounds

5. The thickness of a wallet
   A ounces C feet
   (B) inches D miles

6. The capacity of a spoon
   F ounces H cups
   (G) fluid ounces J gallons

Choose the letter of the best answer.

7. 8 gallons = ■ quarts
   A 2 C 16
   B 4 (D) 32

8. 96 ounces = ■ pounds
   F 1536 H 8
   G 16 (J) 6

9. 9 yards = ■ feet
   A 3 (C) 27
   B 12 D 108

10. 4.5 cups = ■ fluid ounces
    F 2.25 (H) 36
    G 9 J 72

11. 3 miles = ■ feet
    (A) 15,840 C 1584
    B 10,560 D 36

12. 16,000 pounds = ■ tons
    F 4 H 80
    (G) 8 J 1000

13. Awilda has 3 ft of ribbon.
    a. Write this measurement in inches.
    __36 inches__
    b. Suppose Awilda uses 20 in. of her ribbon to wrap a box. How much ribbon does she have left?
    __16__ in., or __1__ ft __4__ in.

LESSON 5-6 Practice B
Customary Measurements

Choose the most appropriate customary unit for each measurement. Justify your answer.

1. the weight of a paperback book
   Possible answer: Ounces; the weight of a paperback book is similar to the weight of several slices of bread, each of which is about 1 oz.

2. the capacity of a large soup pot
   Possible answer: Gallons; the capacity of a large soup pot is greater than the capacity of a large milk jug, which has the capacity of 1 gal.

3. the length of a dining room table
   Possible answer: Feet; the length of a dining room table is similar to the length of 8 or 10 sheets of paper, each of which has a length of about 1 ft.

4. the weight of an elephant
   Possible answer: Tons; the weight of an elephant is similar to the weight of a few buffalos, each of which has a mass of about 1 ton.

Convert each measure.

5. 6 mi to feet   __31,680__
6. 104 oz to pounds   __6.5 lb__
7. 12 qt to pints   __24 pt__
8. 5,000 lb to tons   __2.5 tons__
9. 48 yd to feet   __144 ft__
10. 96 fl oz to pints   __6 pt__
11. 6.5 ft to inches   __78 in.__
12. 20 qt to gallons   __5 gal__
13. $3\frac{1}{4}$ lb to ounces   __52 oz__

14. Marina has 2.5 lb of cashews. She puts 6 oz of cashews in a bag and gives the bag to her brother. What weight of cashews does Marina have left?

$2\frac{1}{8}$ lb, or 2 lb 2 oz

15. Faye is 5 ft 5 in. Faye is 10 in. shorter than her older brother. How tall is Faye's older brother?

6 ft 3 in.

Practice C
5-6 Customary Measurements

Choose the most appropriate customary unit for each measurement. Justify your answer.

1. the weight of an encyclopedia

   Possible answer: Pounds; the weight of an encyclopedia is a few times greater than the weight of a bag of 3 apples, which is about 1 lb.

2. the distance between two cities

   Possible answer: Miles; the distance between two cities is greater than the combined length of 18 football fields, which is about 1 mi.

3. the capacity of a medicine cup

   Possible answer: Fluid ounces; the capacity of a medicine cup is similar to the combined capacity of 2 tablespoons, which is about 1 fl oz.

4. the length of a hamster

   Possible answer: Inches; the length of a hamster is similar to the combined length of several small paper clips, each of which has a length of about 1 in.

Convert each measure.

5. 4.6 tons to pounds — 9200 lb
6. 6600 ft to miles — 1.25 mi
7. 21 qt to pints — 42 pt
8. 5 yd to inches — 180 in.
9. 40 pt to gallons — 5 gal
10. 148 oz to pounds — 9.25 lb

Compare. Write <, >, or =.

11. 25,000 ft < 5 mi
12. 6 lb = 96 oz
13. 7.5 qt > 56 fl oz
14. 4 ft > 32 in.
15. 2 gal < 48 c
16. 4 yd > 136 in.

17. Evan has 1 gal of orange juice. He uses 30 fl oz of orange juice to make a batch of smoothies. How much orange juice does Evan have left?

    98 fl oz

18. Helen has 3.5 yd of fabric. She buys 10 additional feet of fabric. How much fabric does Helen have now?

    20.5 ft, or 20 ft 6 in.

Reteach
5-6 Customary Measurements

You can use the facts in this table to help you convert customary units.

Length	Weight	Capacity
12 in. = 1 ft	16 oz = 1 lb	8 fl oz = 1 c
3 ft. = 1 yd	2,000 lb = 1 ton	2 c = 1 pt
5,280 ft = 1 mi		2 pt = 1 qt
		4 qt = 1 gal

To change larger units to smaller units, multiply.
3 lb = ■ oz
3 × 16 = 48
3 lb = 48 oz

To change smaller units to larger units, divide.
60 in. = ■ ft
60 ÷ 12 = 5
60 in. = 5 ft

1 lb = 16 oz — Multiply the number of pounds by 16.

12 in. = 1 ft — Divide the number of inches by 12.

Convert each measure.

1. 12 qt = ■ gal
   Smaller unit → larger unit
   Fact: __4__ qt = 1 gal
   Operation: divide
   12 ÷ 4 = __3__
   12 qt = __3__ gal

2. 9 yd = ■ ft
   Larger unit → smaller unit
   Fact: 1 yd = __3__ ft
   Operation: multiply
   9 × __3__ = __27__
   9 yd = __27__ ft

3. 7 lb = ■ oz
   larger unit → smaller unit
   7 lb = __112__ oz

4. 48 fl oz = ■ c
   smaller unit → larger unit
   48 fl oz = __6__ c

5. 8 ft = __96__ in.

6. 4,000 lb = __2__ tons

Challenge
5-6 Conversion Diversion

The table shows data about some of the most famous baseball parks in the United States. Use the table to answer the questions.

Baseball Field	Team	Year Opened	Year Closed	Cost	Dimensions of the Center Field Line
Ebbets Field	Brooklyn Dodgers	1913	1957	$750,000	393 ft
Forbes Field	Pittsburgh Pirates	1909	1970	$1 million	400 ft
Tiger Stadium	Detroit Tigers	1912	1999	$8 million	440 ft
Yankee Stadium	New York Yankees	1923		$50 million	408 ft
L.A. Coliseum	Los Angeles Dodgers	1958	1961	$950,000	420 ft

1. How many inches was the center field line in Tiger Stadium?

   5280 in.

2. How many thousands of dollars did it cost to build Yankee Stadium?

   50,000 thousands

3. For about how many months was the L.A. Coliseum open?

   about 36 months

4. How many quarters (25¢) would it have taken to build the L.A. Coliseum?

   3,800,000 quarters

5. How many inches longer was center field line in Forbes Field than in Ebbets Field?

   84 in. longer

6. How many miles to the nearest hundredth of a mile was the center field line in the L.A. Coliseum?

   0.08 mi

7. For about how many months was Forbes Field open?

   about 732 months

8. How many yards is the center field line in Yankee Stadium?

   136 yd

Problem Solving
5-6 Customary Measurements

Write the correct answer.

1. In 2003, a popcorn sculpture of King Kong was displayed in London. The sculpture was 13 ft tall and 8.75 ft wide. How many inches wide was the sculpture?

   105 in.

2. A pilot whale weighs 1500 lb. A walrus weighs 1.6 tons. Which weighs more? How much more?

   The walrus weighs 1,700 lb more.

3. A zoo has a rhesus monkey that weighed 20 lb. The monkey became sick and lost 18 oz. What was the monkey's new weight?

   18.875 lb, or 18 lb 14 oz

4. Two containers have capacities of 192 fl oz and 1.25 gal. Which container has a greater capacity? How much greater?

   The 192 fl oz-container can hold 32 fl oz, or 0.25 gal more.

Choose the letter for the best answer.
This table gives lengths and weights for some apes.

Apes		
Name	Maximum Height	Maximum Weight
Chimpanzee	4 ft	115 lb
Gorilla	67.2 in.	0.2 tons
Orangutan	1.5 yd	3200 oz
Siamang	36 in.	240 oz

7. Which ape has a maximum weight of 200 lb?
   A Chimpanzee
   B Gorilla
   (C) Orangutan
   D Siamang

5. Which ape has the greatest weight?
   A Chimpanzee
   (B) Gorilla
   C Orangutan
   D Siamang

6. Which ape has the least height?
   F Chimpanzee
   G Gorilla
   H Orangutan
   (J) Siamang

8. Which two apes have a 6-inch difference in height?
   F Chimpanzee and gorilla
   G Gorilla and orangutan
   H Siamang and chimpanzee
   (J) Orangutan and chimpanzee

LESSON 5-6 Reading Strategies
Use a Flowchart

This flowchart shows two ways to solve this problem: 56 oz = ▨ pounds.

```
Identify equivalent measures that have the
same units as the measures in the problem.
             16 oz = 1 lb
                │
                ▼
     Choose a method to convert units.
         ╱                    ╲
```

Write a proportion.
Use a ratio of equivalent measures and a ratio of the measures in the problem.

ounces → $\frac{16}{1} = \frac{56}{x}$
pounds →

Solve to find the value of x.
$16 \cdot x = 56 \cdot 1$
$16x = 56$
$x = 3.5$

Write a multiplication equation.
Use a ratio of equivalent measures, which equals 1. Set up the ratio so that you can cancel units.

$56 \text{ oz} = 56 \text{ oz} \times \frac{1 \text{ lb}}{16}$

Simplify. $56 \text{ oz} = \frac{56}{1} \times \frac{1 \text{ lb}}{16}$
$= \frac{56 \text{ lb}}{16}$
$= 3.5 \text{ lb}$

56 oz = 3.5 pounds

Use the problem 18 yd = ▨ ft for Exercises 1-6.

1. What equivalent measures can you use to solve this problem? __1 yd = 3 ft__
2. What proportion can you write to solve this problem? __$\frac{1}{3} = \frac{18}{x}$__
3. Solve the proportion you wrote. __$x = 54$__
4. What multiplication equation can you write to solve this problem?
   $18 \text{ yd} = \frac{18 \text{ yd}}{1} \times \frac{3 \text{ ft}}{1 \text{ yd}}$
5. Simplify the equation you wrote. __18 yd = 54 ft__
6. How do your answers to Exercises 3 and 5 compare?
   __They both show that 18 yd = 54 ft.__

LESSON 5-6 Puzzles, Twisters, & Teasers
Because It's Customary

Choose the most appropriate customary unit for each measurement. Circle the letter above your answer.

1. the capacity of a hot water heater
   B   (T)
   fluid ounces   gallons

2. the weight of a bowling ball
   (D)   K
   pounds   tons

3. the thickness of a book
   (N)   C
   inches   feet

4. the capacity of a thermos
   A   (H)
   gallons   quarts

Next, choose the measurement that is equivalent to the one given. Circle the letter above your answer.

5. 6 ft = ?
   S   (W)
   48 in.   72 in.

6. 12 oz = ?
   (C)   O
   0.75 lb   192 lb

7. 80 fl oz = ?
   (E)   C
   10 c   5 lb

8. 3.5 tons = ?
   P   (O)
   56 lb   7000 lb

9. 36 ft = ?
   (K)   O
   12 yd   3 yd

10. 40 qt = ?
    I   (Y)
    160 gal   10 gal

Write the circled letters above the problem numbers to solve the riddle.

Why do lions eat raw meat?

T H E Y   D O N T   K N O W
1. 4. 7. 10. 2. 8. 3. 1. 9. 3. 8. 5.

H O W   T O   C O O K
4. 8. 8. 1. 8. 6. 8. 8. 9.

LESSON 5-7 Practice A
Similar Figures and Proportions

Identify the corresponding sides.

1. AB corresponds to __XY__.
2. BC corresponds to __YZ__.
3. AC corresponds to __XZ__.

Identify the corresponding sides. Then use ratios to determine whether the triangles are similar. teach

4.
   similar; $\frac{AB}{DE} = \frac{BC}{EF} = \frac{AC}{DF}$;
   $\frac{3}{6} = \frac{4}{8} = \frac{5}{10} = \frac{1}{2}$

5.
   not similar; $\frac{JK}{MN} \neq \frac{KL}{NP}$; $\frac{JL}{MP}$
   $\frac{5}{6} \neq \frac{7}{8}$

6.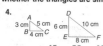
   similar; $\frac{GH}{RS} = \frac{HJ}{ST} = \frac{GJ}{RT}$;
   $\frac{3}{9} = \frac{6}{18} = \frac{5}{15} = \frac{1}{3}$

7.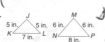
   Possible answer: could be switched
   similar; $\frac{XY}{UV} = \frac{YZ}{VW} = \frac{XZ}{UW}$;
   $\frac{4}{5} = \frac{4}{5}$

Use the properties of similarity to determine whether the figures are similar.

8.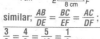
   not similar;
   $\frac{AB}{QR} = \frac{BC}{RS} = \frac{CD}{ST} = \frac{DA}{TQ}$; $\frac{2}{6} \neq \frac{4}{8}$

9.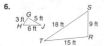
   similar; $\frac{GH}{WX} = \frac{HJ}{XY} = \frac{JK}{YZ} = \frac{GK}{WZ}$;
   $\frac{7}{10.5} = \frac{6}{9} = \frac{3}{4.5} = \frac{2}{3}$

LESSON 5-7 Practice B
Similar Figures and Proportions

Identify the corresponding sides in each pair of triangles. Then use ratios to determine whether the triangles are similar.

1.
   not similar; $\frac{AB}{DE} \neq \frac{BC}{EF}$;
   $\frac{2}{6} \neq \frac{7}{14}$

2. similar; $\frac{RS}{UV} = \frac{ST}{VW} = \frac{RT}{UW}$;
   $\frac{3}{9} = \frac{5}{15} = \frac{1}{3}$

3.
   similar; $\frac{XY}{JL} = \frac{YZ}{JK} = \frac{XZ}{KL}$;
   $\frac{10}{8} = \frac{25}{20} = \frac{30}{24} = \frac{5}{4}$

4. not similar; $\frac{FG}{QR} \neq \frac{EF}{PQ} \neq \frac{EG}{PR}$;
   $\frac{13}{16} \neq \frac{15}{20} \neq \frac{16}{21}$

Use the properties of similarity to determine whether the figures are similar.

5.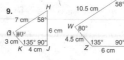
   similar; $\frac{AB}{EF} = \frac{AD}{EH}$
   $\frac{24}{36} = \frac{28}{42} = \frac{2}{3}$

6. not similar; corresponding angles are not equal

LESSON 5-7 Practice C
Similar Figures and Proportions

Use the properties of similarity to determine whether the figures are similar.

1.

    similar; $\dfrac{XY}{RS} = \dfrac{YZ}{ST} = \dfrac{XZ}{RT}$;
    $\dfrac{3.2}{1.6} = \dfrac{6.4}{3.2} = \dfrac{4.8}{2.4} = 2$

2.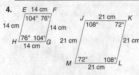

    not similar; $\dfrac{BC}{DE} \neq \dfrac{AB}{EF} \neq \dfrac{AC}{DF}$;
    $\dfrac{1.7}{3.4} \neq \dfrac{2.2}{5.0} \neq \dfrac{2.5}{5.5}$

3.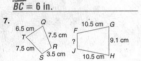

    not similar; $\dfrac{WX}{NO} = \dfrac{ZY}{MP} \neq \dfrac{WZ}{MN} \neq$
    $\dfrac{XY}{OP}$; $\dfrac{12}{16} = \dfrac{6}{8} \neq \dfrac{10}{15}$

4.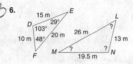

    not similar; angles are not equal

The figures in each pair are similar. Find the missing lengths or angle measures.

5.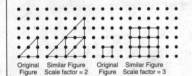

    $\overline{BC} = 6$ in.

6.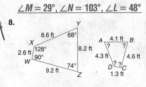

    $\angle M = 29°, \angle N = 103°, \angle L = 48°$

7.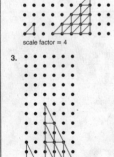

    $\overline{FJ} = 4.9$ cm

8.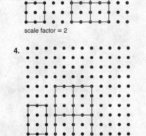

    $\angle A = 68°, \angle B = 74°,$
    $\angle C = 90°, \angle D = 128°$

LESSON 5-7 Reteach
Similar Figures and Proportions

Figures that have the same shape but not the same size are **similar figures**. In similar figures, the ratio of the lengths of the corresponding sides are proportional, and the corresponding angles have equal measures.

To determine if △ABC is similar to △XYZ, you can write a proportion for each pair of corresponding sides.

longest sides	middle sides	shortest sides
$\dfrac{AB}{XY} = \dfrac{15}{10} = \dfrac{3}{2}$	$\dfrac{BC}{YZ} = \dfrac{12}{8} = \dfrac{3}{2}$	$\dfrac{AC}{XZ} = \dfrac{9}{6} = \dfrac{3}{2}$

The corresponding sides are always in the ratio $\dfrac{3}{2}$. So the triangles are similar.

If a polygon has more than 3 sides, you must also show that the corresponding angles are equal.

Identify the corresponding sides. Use ratios to determine whether the figures are similar.

1. $\dfrac{TU}{EF} = \dfrac{16}{8} = \dfrac{2}{1}$, $\dfrac{SU}{DF} = \dfrac{12}{6} = \dfrac{2}{1}$,
    $\dfrac{ST}{DE} = \dfrac{10}{4} = \dfrac{5}{2}$

    Are the ratios proportional? __no__
    Are the triangles similar? __no__

2. $\dfrac{PQ}{FG} = \dfrac{16}{12} = \dfrac{4}{3}$, $\dfrac{PR}{FH} = \dfrac{24}{18} = \dfrac{4}{3}$,
    $\dfrac{QR}{GH} = \dfrac{20}{15} = \dfrac{4}{3}$

    Are the ratios proportional? __yes__
    Are the triangles similar? __yes__

3. similar

4. not similar; angles not equal measure

LESSON 5-7 Challenge
The Same, Only Bigger

You can sometimes create a similar figure by using copies of the original figure.

Notice that the scale factor tells you how many times to repeat the original figure along each side or edge of the similar figure.

Use the given scale factor and copies of the original figure to draw a figure similar to the original figure.

1. scale factor = 4
2. scale factor = 2
3. scale factor = 3
4. scale factor = 2

5. Draw a figure in the space below. Use a scale factor of 2 to create a similar figure. **Drawings will vary. Possible drawing given.**

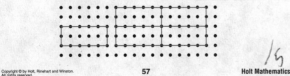

LESSON 5-7 Problem Solving
Similar Figures and Proportions

Use the information in the table to solve problems 1–3.

Painting	Artist	Original Size (in.)
Mona Lisa	Leonardo da Vinci	30 by 21
The Dance Class	Edgar Degas	33 by 30
The Blue Vase	Paul Cézanne	22 by 18

1. A small reproduction of one of the paintings in the list is similar in size. The reproduction measures 11 inches by 10 inches. Of which painting is this a reproduction?

   The Dance Class

2. A local artist painted a reproduction of Cézanne's painting. It measures 88 inches by 72 inches. Is the reproduction similar to the original? What is the ratio of corresponding sides?

   yes; 1:4

3. A poster company made a poster of da Vinci's painting. The poster is 5 feet long and 3.5 feet wide. Is the poster similar to the original Mona Lisa? What is the ratio of corresponding sides?

   yes; 1:2

Choose the letter for the best answer.

4. Triangle ABC has sides of 15 cm, 20 cm, and 25 cm. Which triangle could be similar to triangle ABC?
   - (A) A triangle with sides of 3 cm, 4 cm, and 5 cm
   - B A triangle with sides of 5 cm, 6 cm, and 8 cm
   - C A triangle with sides of 30 cm, 40 cm, and 55 cm
   - D A triangle with sides of 5 cm, 10 cm, and 30 cm

5. A rectangular picture frame is 14 inches long and 4 inches wide. Which dimensions could a similar picture frame have?
   - F Length = 21 in.; width = 8 in.
   - G Length = 35 in.; width = 15 in.
   - (H) Length = 49 in.; width = 14 in.
   - J Length = 7 in.; width = 3 in.

6. A rectangle is 12 meters long and 21 meters wide. Which dimensions correspond to a nonsimilar rectangle?
   - A 4 m; 7 m
   - B 8 m; 14 m
   - C 20 m; 35 m
   - (D) 24 m; 35 m

7. A rectangle is 6 feet long and 15 feet wide. Which dimensions correspond to a similar rectiangle?
   - F 8 ft; 24 ft
   - (G) 10 ft; 25 ft
   - H 15 ft; 35 ft
   - J 18 ft; 40 ft

LESSON 5-7 Reading Strategies
Understanding Vocabulary

Similar means almost the same. If two objects are similar, they have some things in common.

Similar figures are figures that are nearly the same. Similar figures have the same shape, but are different sizes.

Similar figures have **corresponding sides** and **corresponding angles**. *Corresponding* means matching. Each side and angle in a similar figure has a corresponding side and angle.

These two triangles are similar.

Use the figures above to answer each question.

1. What angle corresponds to angle *B*?
   ∠ angle *E*

2. What angle corresponds to angle *A*?
   ∠ angle *D*

3. What side corresponds to side *BC*?
   side $\overline{EF}$

Are the figures similar? Answer yes or no for each pair.

4. yes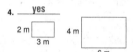

5. no

LESSON 5-7 Puzzles, Twisters & Teasers
Concentrating on Figures

Pretend this is a game of concentration. The object of the game is to match cards with similar figures. Each box represents a card with a figure on it. When you match 2 cards, cross them out. Rearrange the letters of the unmatched cards to solve the riddle.

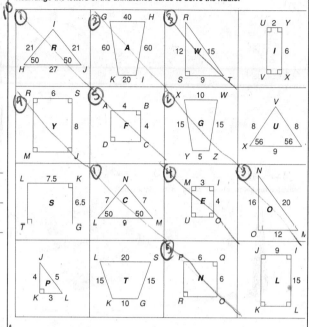

What kind of flowers are on your face?

T U L I P S

LESSON 5-8 Practice A
Using Similar Figures

For each pair of similar figures write a proportion containing the unknown length. Then solve.

1.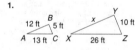

2.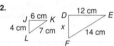

Possible answers are given.

$\frac{10}{5} = \frac{x}{12}$; x = 24 ft

$\frac{14}{7} = \frac{x}{4}$; x = 8 cm

3.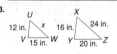

4.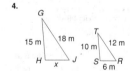

$\frac{12}{16} = \frac{x}{24}$; x = 18 in.

$\frac{18}{12} = \frac{x}{6}$; x = 9 m

5. Kareem and Julio have rectangular model train layouts that are similar to each other. Julio's layout is 4 feet by 7 feet. Kareem's layout is 6 feet wide. What is the length of Kareem's layout?

   10.5 feet

6. A 6-foot-tall adult casts a shadow that is 15 feet long. Estimate the height of a child who casts a 10-foot shadow.

   4 feet

LESSON 5-8 Practice B
Using Similar Figures

△ABC ~ △DEF in each pair. Find the unknown lengths.

1.

x = 60 cm

2.

x = 21 ft

3.

x = 5 m

4.

x = 12 in.

5. The two rectangular picture frames at the right are similar. What is the height of the larger picture frame?

   2.8 feet

   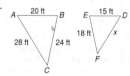

6. A palm tree casts a shadow that is 44 feet long. A 6-foot ladder casts a shadow that is 16 feet long. Use Estimate the height of the palm tree.

   16.5 feet

   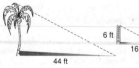

LESSON 5-8 Practice C — Using Similar Figures

Find the unknown length in each pair of similar figures.

1.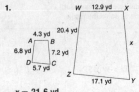

   $x = 21.6$ yd

2.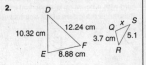

   $x = 4.3$ cm

3.

   $x = 32.5$ ft

4.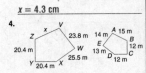

   $x = 22.1$ m —or— 25.5 m

Estimate the height of each object in the picture below.

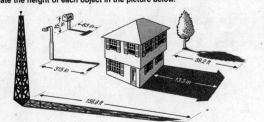

5. house ___9.5 meters___
6. tree ___28 feet___
7. lamppost ___225 inches___
8. radio tower ___112 feet___

LESSON 5-8 Reteach — Using Similar Figures

If you know that 2 figures are similar, you can use proportions to find unknown lengths of sides.

The triangles are similar.
Side AC corresponds to side DF.
Side AB corresponds to side DE.
Side BC corresponds to side EF.

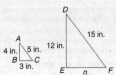

Write a proportion comparing the lengths of a pair of corresponding sides.

$$\frac{AC}{DF} = \frac{BC}{EF}$$
$$\frac{5}{15} = \frac{3}{n}$$
$$5 \cdot n = 15 \cdot 3$$
$$5n = 45$$
$$\frac{5n}{5} = \frac{45}{5}$$
$$n = 9$$

The length of the missing side is 9 in.

Find the unknown length in each pair of similar figures.

1.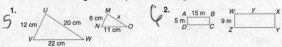

   $\frac{UW}{MO} = \frac{UV}{MN}$; $\frac{20}{x} = \frac{12}{6}$

   $x = 10$ cm

2. $\frac{WZ}{AD} = \frac{WX}{AB}$; $\frac{9}{5} = \frac{y}{15}$

   $y = 27$ m

3. $k = 16$ ft

4. $s = 8$ in.

LESSON 5-8 Challenge — You Be the Artist

Your club is planning a poster to advertise the school's international dinner. The poster will be enlarged and used as a mural on the school cafeteria wall. The poster will also be reduced and used as flyers. The mural will be 10 feet high and 15 feet long. The flyers will be printed on $8\frac{1}{2}$-by-11-inch paper.

Plan the size of the poster so that the enlargement and reduction will be easy to make.

1. What is the width-to-length ratio for the wall mural? Write the ratio in simplest terms.

   10:15; 2:3

2. What is the width-to-length ratio for the flyer? Write the ratio in simplest terms.

   8.5:11; 17:22

3. Do you want your artwork to fill the entire page for the flyer?

   No, you need to leave room to print information about the dinner.

4. What are some possible dimensions for your poster?

   2 feet by 3 feet or 1 foot by 1.5 feet

5. Will your poster fill the wall space when it is enlarged for the mural? Explain.

   Yes, the ratios are equal.

6. What would be a good size for the artwork on the flyer?

   6 inches by 9 inches

LESSON 5-8 Problem Solving — Using Similar Figures

Write the correct answer.

1. An architect is building a model of a tennis court for a new client. On the model, the court is 6 inches wide and 13 inches long. An official tennis court is 36 feet wide. What is the length of a tennis court?

   78 feet long

2. Mr. Hemley stands next to the Illinois Centennial Monument at Logan Square in Chicago and casts a shadow that is 18 feet long. The shadow of the monument is 204 feet long. If Mr. Hemley is 6 feet tall, how tall is the monument?

   68 feet tall

3. The official size of a basketball court in the NBA is 94 feet by 50 feet. The basketball court in the school gym is 47 feet long. How wide must it be to be similar to an NBA court?

   25 feet wide

4. Two rectangular desks are similar. The larger one is 42 inches long and 18 inches wide. The smaller one is 35 inches long. What is the width of the smaller desk?

   15 inches wide

Choose the letter for the best answer.

5. An isosceles triangle has two sides that are equal in length. Isosceles triangle ABC is similar to isosceles triangle XYZ. What proportion would you use to find the length of the third side of triangle XYZ?

   A $\frac{BC}{XZ} = \frac{AB}{XY}$
   C $\frac{AB}{XY} = \frac{AC}{XZ}$
   B $\frac{AC}{XY} = \frac{BC}{XZ}$
   D $\frac{AB}{XY} = \frac{BC}{YZ}$

6. The dining room at Montiocllo, Thomas Jefferson's home in Virginia, is 216 inches by 222 inches. Of the following, which size rug would be similar in shape to the dining room?

   F 72 inches by 74 inches
   G 108 inches by 110 inches
   H 118 inches by 111 inches
   J 84 inches by 96 inches

7. A 9-foot street sign casts a 12-foot shadow. The lamppost next to it casts a 24-foot shadow. How tall is the lamppost?

   A 24 feet
   B 15 feet
   C 18 feet
   D 36 feet

Holt Mathematics

Reading Strategies
5-8 Use a Flowchart

It is very difficult to measure the height of tall tree or a utility pole directly. You can set up proportions to measure very tall objects indirectly.

This method of measuring is called **indirect measurement.** You do not actually measure the height. You use a proportion to find the height.

These two triangles are similar.
Find length x in triangle DEF.

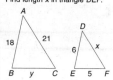

The flowchart helps you set up a proportion to find the value of x.

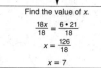

Answer each question.

1. Why is this method is called indirect measurement?

   Possible answer: because you are not actually measuring, but using proportions to find a missing length

2. What is the next step after setting up the proportion?

   Put the lengths of the sides into the proportion.

3. Write a proportion to find the length y in triangle ABC.

   Possible answer: $\frac{18}{6} = \frac{y}{5}$

Puzzles, Twisters & Teasers
5-8 Measure This!

Use indirect measurement to find the width of the smaller rectangle. Show the steps you used by filling in the blanks in the sentences. To solve the riddle, find the letter(s) in each answer with a number below it. Match the letters to the numbered blanks in the riddle.

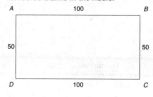

1. Write a P R O P O R T I O N.
          — — — — — — 2 — — — —

2. Substitute the L E N G T H of the sides.
                — — — — 6 —

3. Find the C R O S S   P R O D U C T S.
           — — 4 — —   — — — 3 — — —

4. S O L V E for x.
   — 1 — — —

5. Divide.

6. x = 25

What do you call a cat that swims and has eight legs?

An O   C   T   O   P   U   S   S
    2   4   6   2   5   3   1   1

Practice A
5-9 Scale Drawings and Scale Models

Identify the scale factor. Choose the best answer.

1. Person: 72 inches
   Action figure: 6 inches
   A $\frac{1}{7}$ C $\frac{1}{12}$
   B $\frac{1}{10}$ D $\frac{1}{15}$

2. Dog: 24 inches
   Stuffed animal: 8 inches
   F $\frac{1}{3}$ H $\frac{1}{5}$
   G $\frac{1}{4}$ J $\frac{1}{6}$

3. Fish: 16 inches
   Fishing lure: 2 inches
   A $\frac{1}{6}$ C $\frac{1}{12}$
   B $\frac{1}{8}$ D $\frac{1}{14}$

4. House: 30 feet
   Dollhouse: 3 feet
   F $\frac{1}{3}$ H $\frac{1}{27}$
   G $\frac{1}{10}$ J $\frac{1}{33}$

Identify the scale factor.

5. | | Guitar | Ukulele |
   |---|---|---|
   | Length (in.) | 36 | 18 |

   $\frac{1}{2}$

6. | | Car | Toy Car |
   |---|---|---|
   | Length (ft) | 12 | 3 |

   $\frac{1}{4}$

7. | | Flute | Piccolo |
   |---|---|---|
   | Length (in.) | 30 | 10 |

   $\frac{1}{3}$

8. | | Poodle | Toy Poodle |
   |---|---|---|
   | Height (in.) | 56 | 8 |

   $\frac{1}{7}$

9. On a road map of New York, the distance from New York City to Albany is 3 inches. What is the actual distance between the cities if the map scale is 1 inch = 50 miles?

   150 miles

10. On a scale drawing, a bookshelf is 8 inches tall. The scale factor is $\frac{1}{8}$. What is the height of the bookshelf?

    64 inches

Practice B
5-9 Scale Drawings and Scale Models

Identify the scale factor.

1. | | Alligator | Toy Alligator |
   |---|---|---|
   | Length (in.) | 175 | 7 |

   $\frac{1}{25}$

2. | | Airplane | Model |
   |---|---|---|
   | Length (ft) | 24 | 3 |

   $\frac{1}{8}$

3. | | Car | Toy Car |
   |---|---|---|
   | Length (ft) | 13.5 | 1.5 |

   $\frac{1}{9}$

4. | | Person | Action Figure |
   |---|---|---|
   | Height (in.) | 66 | 6 |

   $\frac{1}{11}$

5. | | Boat | Model |
   |---|---|---|
   | Length (in.) | 128 | 8 |

   $\frac{1}{16}$

6. | | Fish | Fishing Lure |
   |---|---|---|
   | Length (in.) | 18 | 2 |

   $\frac{1}{9}$

7. | | Tiger | Stuffed Animal |
   |---|---|---|
   | Length (in.) | 70 | 14 |

   $\frac{1}{5}$

8. | | House | Dollhouse |
   |---|---|---|
   | Height (ft) | 39.2 | 2.8 |

   $\frac{1}{14}$

9. On a scale drawing, a school is 1.6 feet tall. The scale factor is $\frac{1}{22}$. Find the height of the school.   35.2 feet

10. On a road map of Pennsylvania, the distance from Philadelphia to Washington, D.C. is 0.0 centimeters. What is the actual distance between the cities if the map scale is 2 centimeters = 40 miles?   136 miles

11. On a scale drawing, a bicycle is $6\frac{4}{5}$ inches tall. The scale factor is $\frac{1}{6}$. Find the height of the bicycle. $40\frac{4}{5}$ inches

LESSON 5-9 Practice C
Scale Drawings and Scale Models

Identify the scale factor.

1.
	Bear	Stuffed Animal
Height (in.)	62	15.5

$\frac{1}{4}$

2.
	House	Dollhouse
Height (ft)	32.4	2.7

$\frac{1}{12}$

3.
	Airplane	Model
Length (ft)	25.5	1.5

$\frac{1}{17}$

4.
	Alligator	Toy Alligator
Length (in.)	128.1	6.1

$\frac{1}{21}$

The scale factor of each model is 1:16. Find the missing dimensions.

Item	Actual Dimensions	Model Dimensions
5. barn	height: 32 ft length: 56 ft	height: **2 ft** length: 3.5 ft
6. submarine	length: 300 ft	length: **18¾ ft**
7. bookcase	height: 96 in.	height: **6 in.**
8. tree	height: 40 ft	height: **2½ ft**
9. car	length: 13 ft length: 5.5 ft	length: **9.75 in.** length: **4.125 in.**
10. shark	length: 19 ft	length: **14¼ in.**

11. Hillary took a photograph of her house, which had an actual height of 28.5 feet. If the house measures 3.6 inches tall in the photograph, what is the scale factor?   **1:95**

12. On a road map, the distance from Portland to Seattle is 8 centimeters. What is the actual distance between the cities if the map scale is 2 centimeters = 37.5 miles?   **150 miles**

13. A sculptor plans a statue by making a drawing to scale. On the drawing, the statue is $8\frac{2}{5}$ inches tall. The scale factor in the drawing is $\frac{1}{23}$. Find the height of the statue.   **$193\frac{1}{5}$ inches**

LESSON 5-9 Reteach
Scale Drawings and Scale Models

The dimensions of a scale model or scale drawing are related to the actual dimensions by a *scale factor*. The **scale factor** is a ratio.

The length of a model car is 9 in. →
The length of the actual car is 162 in. → $\frac{9 \text{ in.}}{162 \text{ in.}} = \frac{9 \div 9}{162 \div 9} = \frac{1}{18}$

$\frac{9}{162}$ can be simplified to $\frac{1}{18}$. The scale factor is $\frac{1}{18}$.

If you know the scale factor, you can use a proportion to find the dimensions of an actual object or of a scale model or drawing.

- The scale factor of a model train set is $\frac{1}{87}$. A piece of track in the model train set is 8 in. long. What is the actual length of the track?

$\frac{\text{model length}}{\text{actual length}} = \frac{8}{x}$ $\frac{8}{x} = \frac{1}{87}$ $x = 696$

The actual length of track is 696 inches.

- The distance between 2 cities on a map is 4.5 centimeters. The scale on the map is 1 cm = 40 miles.

$\frac{\text{distance on map}}{\text{actual distance}} = \frac{4.5 \text{ cm}}{x \text{ mi}} = \frac{1 \text{ cm}}{40 \text{ mi}}$ $\frac{4.5}{x} = \frac{1}{40}$ $x = 180$

The actual distance is 180 miles.

Identify the scale factor.

1. Photograph: height 3 in.
   Painting: height 24 in.
   $\frac{\text{photo height}}{\text{painting height}} = \frac{3 \text{ in.}}{24 \text{ in.}} = \frac{1}{8}$

2. Butterfly: wingspan 20 cm
   Silk butterfly: wingspan 4 cm
   $\frac{\text{silk butterfly}}{\text{butterfly}} = \frac{4 \text{ cm}}{20 \text{ cm}} = \frac{1}{5}$

3. On a scale drawing, the scale factor is $\frac{1}{12}$. A plum tree is 7 inches tall on the scale drawing. What is the actual height of the tree?   **84 inches**

4. On a road map, the distance between 2 cities is 2.5 inches. The map scale is 1 inch = 30 miles. What is the actual distance between the cities?   **75 miles**

LESSON 5-9 Challenge
Balls of Sports

Each circle below is a scale drawing of a different type of ball used in a sport.

- Measure the diameter of each circle to the nearest tenth of a centimeter.
- Use the scale to find the actual diameter of the ball to the nearest tenth of a centimeter.
- Use the chart below to find the sport in which the ball is used.

Diameter of Balls Used in Various Sports	
Basketball	24.0 cm
Baseball	7.5 cm
Golf	4.2 cm
Table Tennis	3.8 cm
Tennis	6.4 cm
Volleyball	21.0 cm

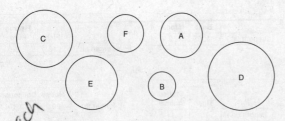

	Circle	Scale	Measured Diameter	Actual Diameter	Sport
1.	A	1 cm = 3 cm	2.5 cm	7.5 cm	baseball
2.	B	1 cm = 15 cm	1.6 cm	24 cm	basketball
3.	C	1 cm = 2 cm	3.2 cm	6.4 cm	tennis
4.	D	1 cm = 1 cm	3.8 cm	3.8 cm	table tennis
5.	E	1 cm = 1.4 cm	3.0 cm	4.2 cm	golf
6.	F	1 cm = 10 cm	2.1 cm	21 cm	volleyball

LESSON 5-9 Problem Solving
Scale Drawings and Scale Models

Write the correct answer.

1. The scale on a road map is 1 centimeter = 500 miles. If the distance on the map between New York City and Memphis is 2.2 centimeters, what is the actual distance between the two cities?   **1,100 miles**

2. There are several different scales in model railroading. Trains designated as O gauge are built to a scale factor of 1:48. To the nearest hundredth of a foot, how long is a model of a 50-foot boxcar in O gauge?   **1.04 feet long**

3. For a school project, LeeAnn is making a model of the Empire State Building. She is using a scale of 1 centimeter = 8 feet. The Empire State Building is 1,252 feet tall. How tall is her model?   **156.5 centimeters tall**

4. A model of the Eiffel Tower that was purchased in a gift shop is 29.55 inches tall. The actual height of the Eiffel Tower is 985 feet, or 11,820 inches. What scale factor was used to make the model?   **1 inch = 400 inches**

Choose the letter for the best answer.

5. The scale factor for Maria's dollhouse furniture is 1:8. If the sofa in Maria's dollhouse is $7\frac{1}{2}$ inches long, how long is the actual sofa?
   A 54 inches   C 84 inches
   B 60 inches   D $15\frac{1}{2}$ inches

6. The Painted Desert is a section of high plateau extending 150 miles in northern Arizona. On a map, the length of this desert is 5 centimeters. What is the map scale?
   F 1 centimeter = 25 miles
   G 5 centimeters = 100 miles
   H 1 centimeter = 30 miles
   J 1 centimeter = 50 miles

7. Josh wants to add a model of a tree to his model railroad layout. How big should the model tree be if the actual tree is 315 inches and the scale factor is 1:90?
   A 395 inches
   B 39.5 inches
   C 35 inches
   D 3.5 inches

8. The scale on a wall map is 1 inch = 55 miles. What is the distance on the map between two cities that are 99 miles apart?
   F 44 inches
   G 1.8 inches
   H 2.5 inches
   J 0.55 inches

LESSON 5-9 **Reading Strategies**
Read a Map

A **scale drawing** has the same shape, but is not the same size, as the object it represents. A map is an example of a scale drawing.
This is a map of a campground. The scale is 1 cm = 10 ft.
To find how far the campground entrance is from the canoe rental office, follow the steps. Use a centimeter ruler to measure.

campsite 1   water

campsite 2

campsite 3

scale 1 cm = 10 feet

Step 1: Measure the distance in centimeters.
→ The distance is 4 centimeters.

Step 2: Set up a proportion using the map scale as one ratio.
→ $\frac{1 \text{ cm}}{10 \text{ ft}} = \frac{4 \text{ cm}}{x \text{ ft}}$

Step 3: Set up cross products. → $1x = 4 \cdot 10$

Step 4: Solve to find the value of x. → $x = 40$

Use the map to answer each question.

1. How many centimeters is Campsite 3 from the water?

   3 centimeters

2. Write a proportion to find the distance from Campsite 3 to the water.
   Possible answer: $\frac{1}{10} = \frac{3}{x}$

3. How many centimeters is Campsite 3 from the canoe rental office?

   5 centimeters

4. Write a proportion to find the distance from Campsite 3 to the canoe rental office.
   $\frac{1}{10} = \frac{5}{x}$

LESSON 5-9 **Puzzles, Twisters & Teasers**
Let's Rock!

Find and circle these words in the word search. Find a word that solves the riddle. Circle it and write it on the line.

scale   model   drawing   factor   ratio
size   actual   dimensions represent proportion

```
V T S P R O P O R T I O N
M N V S B R M K O L J I R
O Y D I M E N S I O N S E
D A X Z R A T I O F F C P
E O P E V O Q W E A A R R
L I K L P S R T U C L E E
A S D F G H M K Q R T E S
A C T U A L B V E Z O E E
N O D R A W I N G T R O N
A J D F A K L M N T S N T
```

What do you do to make a baby sleep on a space ship?

You R O C K E T.